Ice Ages
Past and Future

Jon Erickson

Ice Ages
Past and Future

Jon Erickson

TAB | **TAB BOOKS**
Blue Ridge Summit, PA

FIRST EDITION
FIRST PRINTING

Copyright © 1990 by TAB BOOKS
Printed in the United States of America

Library of Congress Cataloging in Publication Data

Erickson, Jon, 1948-
    Ice ages : past and future / by Jon Erickson.
       p.   cm.
    Includes bibliographical references.
    ISBN 0-8306-8463-8   ISBN 0-8306-3463-0 (pbk.)
    1. Glacial epoch.   I. Title.
QE697.E68   1990
551.7′92—dc20   89-48455
                                                           CIP

TAB BOOKS offers software for sale. For information and a catalog, please contact
TAB Software Department, Blue Ridge Summit, PA 17294-0850.

Questions regarding the content of this book
should be addressed to:

    Reader Inquiry Branch
    TAB BOOKS
    Blue Ridge Summit, PA 17294-0214

Acquisitions Editor: Roland S. Phelps
Technical Editor: Barbara B. Minich
Production: Katherine Brown

Cover photo courtesy of United States Geological Survey

Contents

Acknowledgments

The author wishes to thank the following organizations for providing photographs for this book: the Department of Agriculture Soil Conservation Service, the National Aeronautics and Space Administration (NASA), the National Oceanic and Atmospheric Administration (NOAA), the National Park Service, the U. S. Army Corps of Engineers, the U. S. Coast Guard, the U. S. Department of Energy, the U. S. Navy, and the U. S. Geological Survey (USGS).

Introduction

We are all descendants of the ice age. Periods of glaciation have spanned the whole of human existence for the past 2 million years. The rapid melting of the continental glaciers at the end of the last ice age spurred one of the most dramatic climate changes in the history of the planet. During this interglacial time, people were caught up in a cataclysm of human accomplishment, including the development of agriculture and animal husbandry. Over the past few thousand years, the Earth's climate has been extraordinarily beneficial, and humans have prospered exceedingly well under a benign atmosphere.

Ice ages have dramatically affected life on Earth almost from the very beginning. It is even possible that life itself significantly changed the climate. All living organisms pull carbon dioxide out of the atmosphere and eventually store it in sedimentary rocks within the Earth's crust. If too much carbon dioxide is lost, too much heat escapes out into the atmosphere. This can cause the Earth to cool enough for glacial ice to spread across the land.

In general the reduction of the level of carbon dioxide in the atmosphere has been equalized by the input of carbon dioxide from such events as volcanic eruptions. Man, however, is upsetting the equation by burning fossil fuels and destroying tropical rain forests, both of which release stored carbon dioxide. This energizes the greenhouse effect and causes the Earth to warm. If the warming is significant enough, the polar ice caps eventually melt.

The polar ice caps drive the atmospheric and oceanic circulation systems. Should the ice caps melt, warm tropical waters could circle the globe and make this a very warm, inhospitable planet.

Over the past century, the global sea level has apparently risen upwards of 6 inches, mainly because of the melting of glacial ice. If present warming trends continue, the seas could rise as much as 6 feet by the next century. This would flood coastal plains, coastal cities, and fertile river deltas, where half the human population lives. Delicate wetlands, where many marine species breed and feed, also would be reclaimed by the sea. In addition, more frequent and severe storms would batter coastal areas, adding to the disaster of the higher seas.

The continued melting of the great ice sheets in the polar regions because of higher ocean temperatures and rising sea levels could cause massive amounts of ice to crash into the ocean. This would further raise the sea level and release more ice, which could more than double the area of sea ice and increase correspondingly the amount of sunlight reflected back into space. The cycle would then be complete as this could cause global temperatures to drop enough to initiate another ice age.

The warm interglacial period prior to the last ice age was actually warmer than this one; yet the higher temperatures were unable to halt the advancing glaciers that enveloped one-third of the northern landscape. The present interglacial appears to have just about run its course. The next ice age is already overdue. Its onset depends on how man alters the climate by adding to the greenhouse effect. Perhaps global warming with man's assistance, might be able to hold off the next ice age for a little while longer.

1

Earth Origins

THERE is an incredible amount of ice in the Solar System. Besides the ice that is locked up in the glaciers on Earth, ice is prevalent on Mars and on the moons of the giant gaseous planets Jupiter, Saturn, Uranus, and Neptune. Pluto, which might have been a moon of Neptune that was knocked out of orbit by a collision with a comet, is thought to be a dirty ice ball. Ice is also the primary constituent of the comets.

Much of the ice in the Solar System exists under extremely low temperatures and high pressure, which cause it to exhibit many properties similar to those of rocks. The rings of Saturn (FIG. 1-1) are comprised of chunks of ice about the size of household ice cubes. Scientists still do not understand why they do not simply evaporate into the vacuum of space.

THE BIRTH OF THE SOLAR SYSTEM

Every few decades, a giant star explodes somewhere in the Milky Way Galaxy and becomes a supernova. The first supernova seen from Earth in 400 years was discovered in early 1987. It was believed to have been a blue giant that exploded in the large Magellanic Clouds that reside outside our galaxy. Supernovas spew their contents into the vast emptiness of space to form nebulae (FIG. 1-2) composed mostly of hydrogen and helium, but containing particles of all the known elements.

Certain portions of a nebula are compressed by density waves propagating through the Galaxy. The compression produces a rapid gravitational collapse. If the density is great enough, the heat thus generated will start a thermonuclear reaction in the core and a star will be born.

About two-thirds of the way outward from the center of the Galaxy lies a single medium-size star we call Sol (FIG. 1-3). Single stars are rare in our galaxy because of the manner in which they develop. A single solar nebula fuses, rather than fissioning into two or more nebulae, which is how most stars are formed. Single stars also appear to be the only ones to have planets. The gravitational interaction of two or more stars would not allow planets to form, or if they did, the planets probably would not last very long.

Orbiting around the Sun during its early stages of development was a protoplanetary disk composed of several bands of coarse particles called planetesimals (FIG. 1-4). From the primordial dust grains generated by a supernova, the planetesimals grew by attracting each other with weak electrical and gravitational forces. As they grew, the small rocky chunks swung around the infant Sun in highly elliptical orbits. This resulted in constant collisions between planetesimals, which adhered to each other and formed larger bodies.

When the Sun first flared, it blew the lighter components of the solar nebula outward. The planetesimals that remained closest to it were composed mostly of stony and metallic minerals. These ranged in size from fine sand grains to huge boulders, some of which were over 50 miles wide. Generally, however, most planetesimals were about the size of a pebble. Further away from the Sun, where temperatures were much colder, solid chunks formed of water ice, frozen carbon dioxide, crystalline methane, and ammonia.

(COURTESY OF NASA)

Fig. 1-1. Saturn and its rings from Voyager 1, taken on October 18, 1980.

The presence of a large amount of gas in the early Solar System provided the terrestrial planets with massive atmospheres. The large gravitational compression within these atmospheres created surface temperatures higher than the melting point of rocks. The formative Sun at first glowed red hot as gases compressed the core. Finally, when the core reached a critical temperature of about 15 million degrees Celsius, the Sun ignited. When hydrogen fused into helium on the surface of the core, copious amounts of energy were produced. The ignition of the Sun created a strong

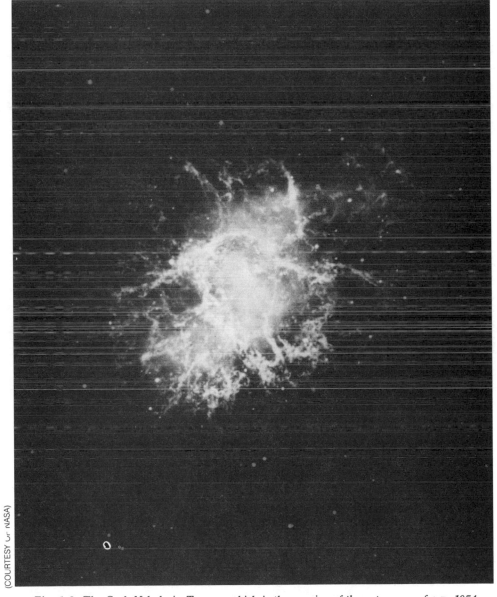

(COURTESY OF NASA)

Fig. 1-2. The Crab Nebula in Taurus, which is the remains of the supernova of A.D. *1054.*

solar wind that was composed of subatomic particles. These eroded away the original atmosheres of the terrestrial planets Mercury, Venus, Earth, and Mars and blew the gases and water vapor outward toward the giant planets, which incorporated them into their atmospheres.

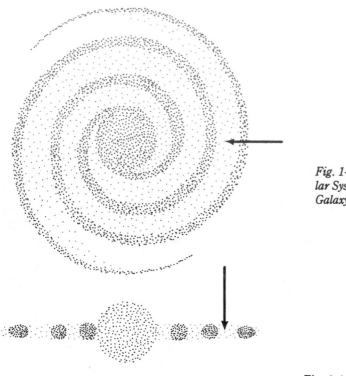

Fig. 1-3. Location of the Solar System in the Milky Way Galaxy.

Fig. 1-4. Formation of the planets from planetesimals.

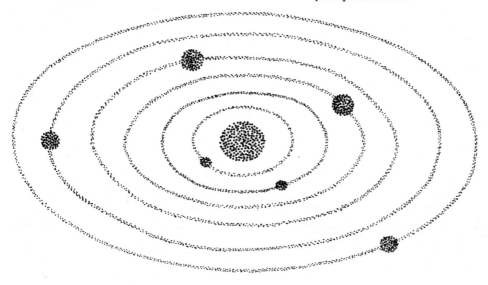

The outer planets probably have rocky cores about the size of the Earth or larger, a mantle probably composed of water and ice, and a thick layer of compressed gas that is mostly hydrogen and helium with smaller amounts of methane and ammonia. Pluto, the moons of the outer planets, and the comets are essentially rocks encased in a thick layer of ice. Jupiter has about the same composition as the Sun. If it had continued growing, it might have ignited like a small companion star, and the Solar System would have resembled many twin-star systems that pervade the Galaxy.

Mars is the closest planet to Earth in relation to its length of day and the tilt of its axis. It has seasons just like those on Earth. It also has two polar ice caps. The South Pole of Mars is capped with permanent carbon dioxide ice, which contains about 20 percent of the planet's atmospheric carbon dioxide during the winter. The South Pole is much colder than the North Pole because the planet is about 25 million miles further away from the Sun during the southern winter than it is during the northern winter. As a result, the South Pole receives 40 percent less heat from the Sun in winter than does the North Pole.

The carbon dioxide in Mars' atmosphere originated from giant volcanoes (FIG. 1-5), which have long ceased erupting. The carbon dioxide ice on Mars' North Pole retreats every summer, and exposes an underlying cap of water ice. During the winter, the ice cap grows halfway to the equator. During the summer, it retreats to a small island of ice at the top of the pole. Combined with the ice buried in the ground, there appears to be a substantial amount of water on Mars. Should Mars warm up again someday, seas several hundred feet deep could cover nearly the entire planet.

THE STRUCTURE OF THE EARTH

Magnetic rocks in the Earth have been dated as old as 2.7 billion years. So, it would not be unexpected for the Earth to have had a molten outer core comparable to its present size at an early age. There are two basic, opposing theories that deal with the origin of the Earth. In the first theory, nebular material of both stony and metallic planetesimals was mixed until the mass grew to nearly the present size of the planet. Then, short-lived, highly radioactive elements called radionuclides heated the interior of the Earth as though it were a giant nuclear reactor.

The Earth's interior began a gradual meltdown because more heat was generated than could escape to the surface. At the same time, a large bombardment of meteorites heated and eventually melted the surface of the planet by impact friction. The Earth thus became a large molten sphere. The denser matrials, particularly the abundant elements of iron and nickel, fell inward toward the center, while the lighter elements rose to the surface.

After about a half billion years, when the short-lived radionuclides had expended their energy, the Earth began to gradually cool down. Long-lived radionuclides, principally uranium, thorium, and potassium, continued to maintain the Earth's interior at a somewhat steady temperature. This allowed convection currents to flow in the core and in the mantle, the intermediate layer of rocks that lies outside the core. At this time there was still no permanent crust, so the first 700 million years of geologic history is not known.

The second theory contends that the core formed first and the mantle and surface rocks were gradually deposited on top of it. The difficulty with this theory is explaining how small particles in the solar nebula stuck together because gravitational forces were extremely weak. When a collision did occur, the particles would simply have bounced off each other or shattered on impact. Particles composed of iron and nickel, however, could easily have clumped together because of their mutual magnetic attraction.

The magnetic particles might have come from the permanent magnetic core of one or more supernovas, which provided the original cosmic material for the solar nebula. As the core grew its large density created enough gravity to attract rocky material out of the preplanitary disk. As more material was added to the core, the force of gravity became greater, until all the planetesimals in the Earth's path were attracted to the core.

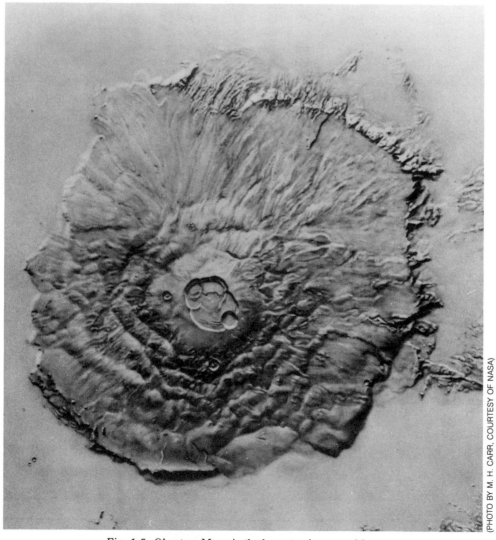

(PHOTO BY M. H. CARR, COURTESY OF NASA)

Fig. 1-5. Olympus Mons is the largest volcano on Mars.

The Earth's core is not permanently magnetized. Any residual magnetism would have been lost because of the intense heat that occurred during the core's separation into a liquid outer layer and a solid inner sphere. When a magnet, such as the Earth's core, is heated above a certain temperature, called the Currie point, its iron atoms vibrate wildly, lose their magnetic alignment, and remain disoriented even after cooling.

The core of the Earth is about 4,300 miles in diameter, or a little over half the diameter of the planet. The inner core is composed of iron-nickel silicates and is roughly 1,500 miles in diameter. The liquid outer core is composed mostly of iron and some nickel. It flows as easily as molten iron ore.

The Earth's core possesses about one-sixth the planet's volume and about one-third its mass. The core is not a smooth sphere, but has ridges as tall as Mount Everest and grooves as deep as the Grand Canyon (FIG. 1 6). This is due to convection currents in the mantle that either press down or pull up the surface of the core through the flowage of mantle rocks. When the convection currents in the mantle slowed down because of the loss of radiogenic heat, lighter rock materials were allowed to migrate toward the surface where they formed a basaltic crust similar to the film on top of a bowl of cold pudding.

Fig. 1-6. The core has bumps and grooves created by convection currents in the overlying mantle. Topography shown is highly exaggerated.

This new crust was peppered by a massive meteorite shower between 4.2 and 3.9 billion years ago. Because the Earth had not yet formed a permanent crust, the meteors simply splashed into the planet, kicking up huge quantities of partially solidified and molten rock. The scars in the crust were quickly healed over with great floods of fresh basalt lava that bled through giant fissures on the surface. When the permanent crust formed, intense weather systems quickly eroded any remaining craters, which is why no telltale signs of the great meteor bombardment can be found on Earth today.

Fig. 1-7. Location of continental shields, the nucleus upon which the continents grew.

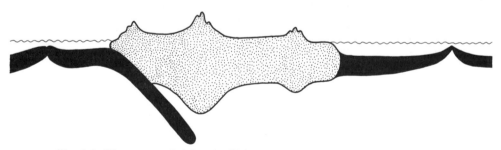

Fig. 1-8. The crust underlying the United States and adjoining oceanic plates.

Up until about 4 billion years ago, no landmasses marred the Earth's watery face. A giant impact might have triggered the evolution of ancient continental shields (FIG. 1-7), upon which the continents grew. No other body in the Solar System has continents like those on Earth. The continental crust (FIG. 1-8) has an average thickness of 20 to 25 miles. In the vicinity of major mountain ranges, however, it reaches a thickness of 45 miles because like an iceberg most of the mountain resides beneath the surface.

By comparison, the oceanic crust is considerably thinner. In most places it is only 3- to 5-miles thick. The continental crust is 20 times older than the oceanic crust, which has been dated to 180 million years. This is because the oceanic crust is consumed by the mantle at subduction zones. Perhaps as many as 20 oceans have come and gone over the last 2 billion years.

A dozen or so lithospheric plates (FIG. 1-9) that act like rafts on a sea of molten rock carry around the separate pieces of crust. The lithosphere, which is the solid portion of the upper mantle, ranges in thickness from 6 to 90 miles, with an average thickness of about 60 miles. The plates spread apart at midocean ridges and converge at deep-sea trenches, where they subduct into the mantle and melt. The plates and oceanic crust are constantly being recycled through the mantle, while the lighter, more buoyant continental crust remains on the surface. This is why the Earth has not been completely covered by water, except perhaps at its earliest formative stages.

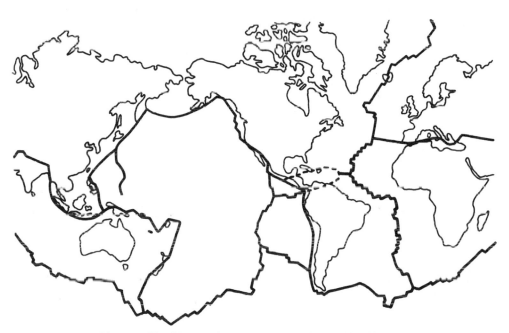

Fig. 1-9. The lithospheric plates that comprise the Earth's crust.

THE CREATION OF THE MOON

One popular theory for the origin of the Moon contends that our planetary companion is from the Earth. Sometime between the Earth's formation and the beginning of the great meteorite bombardment, Jupiter's gravity pulled an asteroid about the size of Mars out of the asteroid belt that lies between Mars and Jupiter. The asteroid grazed the Earth while the planet was still in a molten state. The asteroid's gravitational attraction created a huge tidal bulge on the Earth's surface and sucked a substantial amount of molten rock out of the Earth. As the asteroid sped back into space, material from both the asteroid and the Earth instantaneously vaporized in a mighty explosion. The debris was sent into orbit around the Earth and formed a prelunar disk (FIG. 1-10).

The Moon grew by acquiring rocky material that accumulated in its orbit. After several million years, all the debris in the Moon's eccentric path around the Earth were pulled to it. In the process, radioactivity, compression, and impact friction heated the Moon, so that it became a molten sphere orbiting around the Earth. The result was a satellite with a mass about one-eightieth and a volume about one-fiftieth that of the Earth.

The Moon quickly cooled and formed a permanent crust long before the process was completed on Earth. This is because its ratio of surface area to volume is much greater. Convection in the Moon's mantle and molten iron core might have generated a weak magnetic field early on, but convective motions were not powerful enough to drive plate tectonics as they do on Earth. During the great meteor bombardment, the Moon was intensely cratered. Because it has no atmosphere to produce erosion the Moon still retains much of its original terrain features (FIG. 1-11).

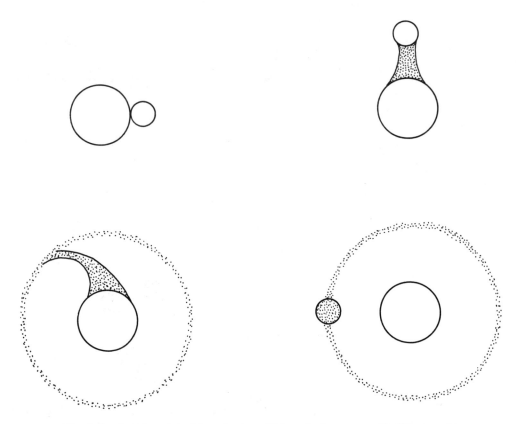

Fig. 1-10. Creation of the Moon by the collision of a large asteroid. The asteroid glanced off the Earth, sucking out a major portion of the planet's interior, which went into orbit and coalesced into the Moon.

Some of the evidence for this theory of lunar evolution comes from the Moon itself. Moon rocks brought back by Apollo astronauts in the late 1960s and early 1970s are believed to be similar in composition to the Earth's upper mantle and range in age from 3.2 to 4.5 billion years. Because no rocks were found that were younger than 3.2 billion years the Moon probably ceased all geologic activity by this time.

The oldest Moon rocks, called Genesis Rock, formed the lunar crust and originated deep within the Moon's interior. The slight variation in composition between the lunar rocks and those of the Earth's mantle might arise from the fact that the Moon has been geologically dead for over 3 billion years, while the rocks in the Earth's mantle have been continuously recycled by convection currents, causing them to change composition over time.

After its formation, the Moon's orbit was so close to the Earth that it filled much of the sky and caused huge tides in the Earth's thin, elastic crust. The Earth's rotation rate was much faster too—days were only a few hours long. The rotation rate decreased as the Earth transferred a portion of its angular momentum, or rotational energy, to the Moon. At the same time, the Moon moved progressively further away. Even today, the Moon is still drifting away from the Earth.

(COURTESY OF NASA)

Fig. 1-11. Photograph of the lunar surface taken by Apollo 17 astronauts.

THE ATMOSPHERE AND OCEAN

Icy visitors from outer space pounded the early Earth and provided it with a substantial amount of water vapor and gases. The barrage of meteorites and comets started around 4.2 billion years ago and lasted for almost a half billion years. Some of these cosmic invaders were composed of stony rock; some were composed of metallic iron and nickel; others were composed of water ice and frozen gases. Comets, which are essentially a rocky core encased in ice, came from the Oort Cloud in the outer reaches of the Solar System. As the atmospheric pressure continued to rise, the smaller meteors burned up on entry due to air friction in the upper atmosphere.

The early atmosphere kept the planet from freezing, even though the Sun's output was about a third less than it is today. The Earth received only as much sunlight as Mars does now. There is no geologic evidence that indicates the Earth was ever frozen over entirely like the moons of Jupiter, Saturn, and Neptune. If the Earth had completely frozen over, it would have remained that way indefinitely because ice reflects sunlight so well. The Sun's output would have had to have increased by 50 percent in order to trigger a global thaw.

Because the Earth recycles rocks in its interior carbon dioxide is continuously being regenerated by volcanic outgassing. Greenhouse gases such as carbon dioxide have the ability to trap incoming radiant energy that would otherwise escape into space. So even though the Sun was weaker, the Earth remained reasonably warm, allowing life to flourish early in its history.

The early atmosphere might have contained as much as 1,000 times more carbon dioxide than it does now. This still accounts for only a tiny fraction of the total amount of carbon stored in the crust in the form of carbonate rocks such as limestone. When the Sun heated up, large amounts of carbon dioxide were scrubbed out of the atmosphere by vigorous weathering processes. This decreased the greenhouse effect and kept the Earth from becoming as hot as Venus, whose dense carbon dioxide atmosphere and thick acid clouds (FIG. 1-12) trap solar radiation. In effect, carbon dioxide acts like a global thermostat. It keeps the Earth's temperature between the freezing and boiling points of water, the range within which life can exist.

Other gases in the atmosphere, such as ammonia, methane, and water vapor, are also good greenhouse gases and might have contributed substantially to the warming of the early Earth. Ammonia in the atmosphere keeps the Sun from breaking down all molecules into nitrogen, which has no effect on greenhouse warming, and hydrogen, which escapes into space due to its light mass.

To counteract the greenhouse effect, an abundant amount of water vapor in the atmosphere produces thick clouds that reflect much of the Sun's energy back into space. Presently, clouds return to space about 30 percent of the sunlight that reaches the upper atmosphere. They also play an important role in regulating the Earth's temperature. When it gets too hot, more water evaporates from the ocean and forms clouds, which reflect sunlight. When it gets too cold, less water evaporates, less clouds form, and more sunlight is allowed to reach the surface. Snow and ice are as effective as clouds in reflecting sunlight back into space.

The primordial atmosphere did not possess significant amounts of free oxygen. Oxygen produced by the breakdown of water vapor into elemental oxygen and hydrogen by ultraviolet light from the Sun quickly combined with metals in the crust to form metal oxides. The hydrogen escaped into space or combined with carbon to make hydrocarbons. The oxygen also recombined with hydrogen and carbon monoxide to form water vapor and carbon dioxide.

Some oxygen also might have reached into the upper atmosphere to create a thin ozone screen that would have helped reduce the breakdown of water vapor by ultraviolet radiation. Sulfur from erupting volcanoes might have worked just as well in shielding the Earth from ultraviolet radiation. This would have prevented the total loss of the Earth's oceans, as might have happened on Venus during its early years.

When the Earth's atmosphere first formed, it held almost enough water to fill an entire ocean, whereas today only about 1 percent of the planet's water is held in the atmosphere. In effect, steam made up a large portion of the early atmosphere. The atmospheric pressure was several times higher than it is today. When the atmosphere cooled enough to allow water vapor to condense into rain, the water evaporated before it reached the ground because the Earth's surface was still very hot. Eventually, when the ground cooled enough to allow rain to fall on it, the evaporation of rainwater further cooled the surface. This initiated a continuous deluge that covered the entire planet in a global sea that was nearly 2 miles deep.

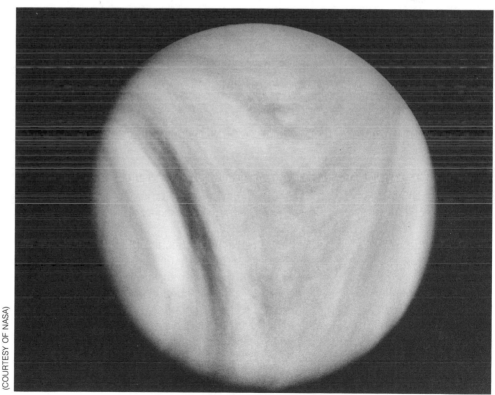

(COURTESY OF NASA)

*Fig. 1-12. A full-disk view of Venus
taken by the Pioneer-Venus Orbitor on February 10, 1979.*

The water vapor reacted chemically with sulfur dioxide and nitrous oxide in the atmosphere to produce acid rain, which in turn reacted with metallic minerals in the crust to create metallic salts. These were carried in solution by streams that emptied into the ocean, and dumped considerable amounts of salt and other chemicals into the sea water. In essence, the ocean became a vast chemical plant that manufactured a huge variety of chemical substances. Rather than slowly accumulating salts and other chemical compounds over a lengthy period of time, the ocean achieved chemical equilibrium early in its history. This might have been responsible for the early emergence of life.

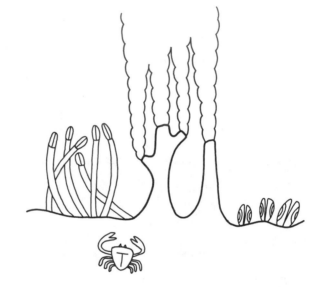

Fig. 1-13. Tall tube worms, giant clams, and large crabs live near hydrothermal vents on the deep ocean floor.

THE ORIGIN OF LIFE

The Earth's primordial atmosphere contained carbon dioxide, ammonia, methane, nitrogen, and water vapor. These gases created a reducing atmosphere, meaning that chemical reactions took place in the absence of oxygen. This was fortunate. Any developing life forms would have been destroyed if substantial amounts of oxygen were present because oxygen interferes with the formation of organic molecules.

The absence of oxygen also allowed the Sun's strong ultraviolet light to reach the lower atmosphere, where it could trigger the production of organic compounds. In addition, the Sun's strong magnetic fields set up an unbalanced electrical charge distribution between the ionosphere (the electrified portion of the upper atmosphere) and the surface of the Earth. This caused powerful lightning bolts to dart to and fro, which initiated additional chemical reactions.

As organic molecules formed, they were weathered out by heavy rains. The rainwater returned to the sea in streams, carrying carbon compounds of every description. These included organic molecules formed in the atmosphere as well as those leached out of the ground. In the ocean, chemical reactions among carbon compounds produced a huge variety of hydrocarbon chains. The carbon chains strung together to form molecules of hydrogen cyanide, ethane, ethylene, and formaldchyde—some of life's first steps. A short time after the formation of the ocean, all the essential amino acids and nucleotides, which are the building blocks of life, were present. The ocean was full of organic substances with the consistency of bouillon.

Life probably had a very difficult time at first because the Earth was constantly being showered with comets and meteorites. The first living cells might have been exterminated repeatedly, forcing life to originate over and over again. As primitive organic molecules attempted to arrange themselves into living cells, frequent impacts might have blasted them apart before they could reproduce. Perhaps the only safe place for life to evolve was on the bottom of the deep sea near hydrothermal vents, where the most bizarre creatures found on Earth currently live (FIG. 1-13).

Discovery of Ice

DOWN through the ages, prophets from many cultures have foretold of the Earth being destroyed either by a great flood or fire, but nothing has ever been mentioned about ice. Only the Norsemen, having experienced the cold of the north, envisioned a time of endless winters when the seas would freeze solid. Perhaps tales of such a disaster were provoked by memories of the great ice age, when one-third of the Earth's land surface was covered by thick sheets of ice. It has only been within the last two centuries that scientists have begun to understand the geologic clues that lead to the discovery of worldwide glaciation.

The study of science started with the early Greeks, who developed a new awareness of the Earth. To further understand the Earth's secrets, the Greeks set about measuring the planet, determining the distance to the Moon, and exploring unknown lands. Although many assertions about natural phenomena made by the great Greek philosopher Aristotle have since been laid to rest, he at least directed the inquisitive minds of science down the right path. It was unfortunate that mankind had to pull itself out of the Dark Ages before continuing to explore the Earth and to discover the new worlds that lay beyond.

THE BIRTH OF GEOLOGY

During the late Renaissance period, there was a rebirth of scientific enquiry into natural phenomena. Around the beginning of the seventeenth century, the French chemist Lemery noticed that a mixture of iron filings, sulfur, and water spontaneously combusted and gave off steam and other projected matter. This led Lemery to the conclusion that volcanoes were produced by the fermentation and combustion of certain matter when they came in contact with air and water. His ideas gave rise to the theory of neptunism, named for Neptune, the Roman god of the sea. The German geologist Abraham Werner expanded on this idea. He maintained that when pyrite was exposed to water, it heated up and ignited coal, which in turn melted the rocks nearby and caused them to erupt.

The Scottish geologist James Hutton argued against the theory of neptunism and advanced an opposing theory called plutonism, named for Pluto, the Greek god of the underworld. Plutonism was based on the premise that the depths of the Earth were in constant turmoil, causing molten matter to rise to the surface through faults and fissures, thus giving rise to an erupting volcano.

The Italian physician Nicolas Steno recognized that in a sequence of undeformed layered rocks, each layer was formed after the one below it and before the one above it. Steno's law of superposition might seem obvious to us today, but in his time it was hailed as an important scientific discovery. Steno also put forward the principle of original horizontality, which states that sedimentary rocks were initially laid down horizontally in the ocean, then subsequent folding and faulting raised them upward out of the sea and inclined them at steep angles.

Although the existence of fossils had been known down through the ages, their significance as a geologic tool was not discovered until the eighteenth century. The English engineer and geologist William Smith, while building canals across Great Britain, found that layers of rock contained fossils that were unlike those in beds either above or below. He also noticed that sedimentary strata in widely separated areas could be identified by their distinctive fossil content. Therefore, geologic formations could be dated relative to each other according to the types of fossils they contained.

In Scandinavia, marine fossil beds were found that had risen more than 1,000 feet above sea level since the last ice age. The weight of the ice sheets depressed the land during the time the marine deposits were laid down. When the ice sheets melted, the removal of the weight caused the land to rise up again due to the principle of isostacy (FIG. 2-1), which is responsible for maintaining equilibrium in the Earth's crust. A similar circumstance has occurred in the Baltic region. Over the centuries, mooring lines on harbor walls have risen so far above the sea that they can no longer be used to tie up ships.

James Hutton, known today as the father of geology, advanced the theory of uniformitarianism (also called gradualism) in 1785. Simply stated, it means that the present is the key to the past. Or, in other words, the forces that shaped the Earth are uniform and operate in the same manner and at the same rate today as they did in the past. Hutton also concluded that granite erratic boulders that rested on top of the limestone rock of the Jura Mountains in Switzerland were carried there by immense glaciers. This seemed to indicate that the Earth was shaped from time to time by certain catastrophic events.

Early in the nineteenth century, the French anatomist Georges Cuvier suggested that the past had experienced a series of environmental catastrophes. In his view, such disruptions were responsible for the disappearance of the mammoths at the end of the last ice age, the large primitive mammals that existed before the ice age, and the giant reptiles, now known as dinosaurs, that became extinct at the end of the Mesozoic era.

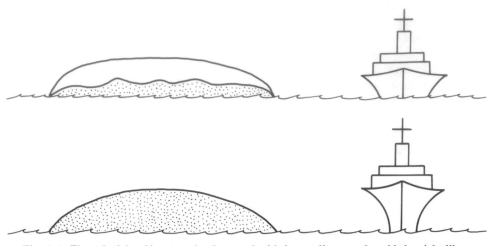

Fig. 2-1. The principle of isostacy: land covered with ice readjusts to the added weight like a loaded freighter. When the ice melts, the land is buoyed upward as the weight lessens.

Seventeenth and eighteenth century geologists spoke of a great flood and argued that it was so devastating it broke apart old continents and created new ones. Taking this idea one step further, the nineteenth century German naturalist and explorer Alexander von Humboldt thought that a huge tidal wave had surged across the globe and carved out the Atlantic Ocean like a giant river valley, leaving the continents divided with opposing shorelines. Von Humboldt was also among the few scientists of his day who accepted the idea that the northern lands were once covered by thick glaciers.

The Austrian geologist Edward Suess made the first serious investigation into why the continents seemed to fit so well together (FIG. 2-2). Suess showed how the

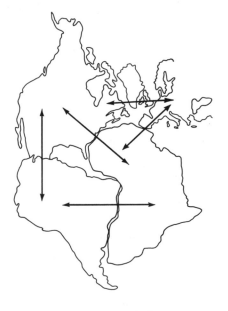

Fig. 2-2. The fitting together of the continents and possible migration routes of plants and animals.

continents of the Southern Hemisphere fit together into a composite landmass he called Gondwanaland, named after a province in India. He called the northern land-mass Laurasia, named for the Canadian Laurentian province and Asia.

By the beginning of the twentieth century, scientists still held the view that land bridges connected the continents, which allowed the migration of plants and animals from one landmass to another, especially when sea levels dropped during times of glaciation. In 1908, the American geologists Frank Taylor and Howard Baker, working independently, suggested an alternative explanation based on continental movements. Taylor suggested that two supercontinents, located at each of the poles, slowly drifted toward the equator. Ancient glacial deposits are found on all the southern continents, indicating that they had once existed together over the South Pole.

On one of his expeditions to Greenland, the early twentieth century German meteorologist and arctic explorer Alfred Wegener might have recognized the similarity between the rifting and drifting of pack ice and the motions of the continents. When an ice floe broke apart, new ice filled the gap. When the new ice froze, it pushed the two halves further away from each other. When two ice floes pressed against each other because of the confined space, chunks of ice broke off and were heaved up into ice ridges. Wegener, however, incorrectly thought that the continents penetrated the ocean floor, which would have caused the edges of the continents to buckle upward and form mountain ranges.

Wegener noticed a high degree of correspondence between the shapes of continental coastlines on either side of the Atlantic Ocean and a similarity between fossils in South America and Africa. He also found matching mountain chains with similar rock formations on the opposing continents, and even ancient climatic conditions that were much the same. It was Wegener's contention that 200 million years ago all the landmasses existed in a single large continent he called Pangaea (FIG. 2-3), which meant all lands. The rest of the world was covered by a single great ocean he called Panthalassa, which meant universal sea.

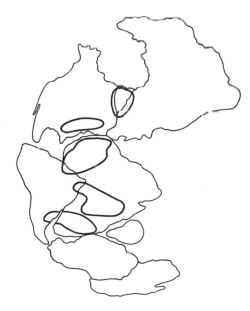

Fig. 2-3. Pangaea showing matching geological provinces.

During his lifetime, Wegener's ideas were considered preposterous and he was ridiculed by the scientific community. Recently, however, scientists have built upon his theory after finding unrefutable evidence for sea floor spreading and the subduction of lithospheric plates into the mantle. This theory is called plate tectonics. The position of the continents and the creation of land masses can now be explained through the interaction of crustal plates. When two plates collide, mountains rise up. When one plate overrides another, volcanoes erupt. When two plates slide past one another, earthquakes shatter the land.

During the 1920s, the Yugoslav geophysicist Milutin Milankovitch painstakingly calculated the changes in incoming solar radiation for every latitude during all seasons. When his labors were completed, he found three orbital cycles that coincided with the 100,000-, 41,000-, and 22,000-year ice age cycles. In 1941, Milankovitch proposed the theory of orbital variations (discussed in more detail in chapter 7), in which he stated that cool summers and not necessarily severe winters were all that were required to trigger an ice age. Unfortunately, like his contemporary Alfred Wegener, his ideas did not gain acceptance by the scientific community until sometime after his death.

ERRATIC BOULDERS

Huge blocks of granite weighing upwards of 20,000 tons, called erratic glacial boulders (FIG. 2-4), have been found strewn across the slopes of the Jura Mountains in Switzerland. Geologists have been able to trace the boulders back to the Swiss Alps over 50 miles away. Most geologists of the late eighteenth century thought these boulders were swept there by the Great Flood.

(PHOTO BY G. K. GILBERT, COURTESY OF USGS)

Fig. 2-4. Erratic glacial boulder in the Sierra Nevada Mountains of California.

In 1760, however, the Swiss geologist Horace de Saussure noticed that the surfaces of projecting rocks downstream from a glacier on the glacial valley floor looked strikingly different from those up on the sides of the valley (FIG. 2-5). The higher rocks were rough and jagged, while the lower rocks were rounded, smooth, and covered with numerous parallel scratches that pointed down the valley. Everywhere, rocks and boulders lay scattered about as though they simply had been dumped there. From this observation, de Saussure concluded that glaciers had once extended far into the valley, and had ground the rocks on the valley floor as the ice advanced and receded.

(PHOTO BY H. E. MALDE, COURTESY OF USGS)

Fig. 2-5. Saskatchewan Glacier in Alberta, Canada, showing eroded glacial valley.

In 1795, the Scottish geologist James Hutton described the Alps as once having been covered by a mass of ice with immense glaciers that carried blocks of granite for great distances. Most geologists at that time, however, refused to believe that a river of solid ice with imbedded rocks had moved along the valley floor like a giant file and ground down the rocks as it flowed over them. Nor would they admit that the glaciers were widespread, and able to drop isolated blocks of granite in the most unlikely places.

The Swiss civil engineer Ignatz Venetz, having heard about the marks left on valley floors by advancing glaciers, visited several glaciers in various parts of the Swiss Alps. By 1829, he had amassed enough information to declare that alpine glaciers had not only covered the Jura Mountains, but had extended far onto the European plain.

The Swiss naturalist Luis Agassiz became the foremost proponent of the glacial theory and declared that glacial ice masses had once blanketed the Swiss Mountains and covered the northern parts of Europe, America, and Asia. In 1837, he led an expedition of the most prestigious geologists in Europe to the Jura Mountains. On the valley floor, they saw large areas of polished and deeply furrowed rocks miles from existing glaciers. Heaped rocks, called moraines, marked the extent of former glaciers (FIG. 2-6). The valley in which the scientists stood was once buried in ice a mile or more thick. The glaciers descended from the mountains, spread across most of northern Europe, and like giant bulldozers destroyed everything in their paths.

Fig. 2-6. Terminal moraine at the margin of a glacier in Deschutes County, Oregon.

Agassiz demonstrated how glaciers formed from the slow accumulation of snow in the mountains. The snow crystals on the bottom were packed and compressed into solid ice by the weight of the overlying snow. After a great many years, a thick solid sea of ice formed. The ice was viscous. It flowed outward or retreated back toward the mountains, depending on the prevailing climate. Agassiz discovered further evidence of massive glaciation in the British Isles, Scandinavia, and the north European plain.

American geologists gladly accepted Agassiz's views, for now they could explain such phenomena as gravel deposits, known as terminal and lateral moraines, and polished and striated rocks (FIG. 2-7). Long Island and Cape Cod were composed of thick moraines. Rocks in western New York State were found to be polished and striated by glaciers just like the rocks of the Jura Mountains. When Agassiz immigrated to America, he found that virtually all of the North American continent north of the Ohio and Missouri rivers had once been glaciated (FIG. 2-8). But most geologist of his day could not imagine ice sheets of the magnitude that Agassiz described.

(PHOTO BY G. K. GILBERT, COURTESY OF USGS)

Fig. 2-7. Polished and striated rocks near Cathedral Lake, Yosemite National Park, California.

Fig. 2-8. Extent of Pleistocene glaciation in the U.S.

THE ICE AGES

By the late 1800s, geologists began to accept Agassiz's glacial theories and discovered there was more than a single episode of glaciation. The study of successive layers of glacial clay separated by soil or peat suggested that several ice ages followed one after another (TABLE 2-1). In 1909, the Swiss geologists Albrecht Penck and Edward Brukner confirmed that there were at least four separate ice ages. The ice ages in the Alps got their names from the Bavarian streams that exposed traces of particular episodes of glaciation. These were named, from the oldest to the youngest, the Gunz, Mindel, Riss, and Wurm ice ages. In America, the corresponding episodes of glaciation were given the names of the affected states and were called the Nebraskan, Kansan, Illinoian, and Wisconsin ice ages (FIG. 2-9).

TABLE 2-1. Chronology of the Major Ice Ages

TIME IN YEARS	EVENT
2 billion	First major ice age.
700 million	The great Precambrian ice age.
230 million	The great Permian ice age.
230–65 million	Interval of warm and relatively uniform climate.
65 million	Climate deteriorates, poles become much colder.
30 million	First major glacial episode in Antarctica.
15 million	Second major glacial episode in Antarctica.
4 million	Ice covers the Arctic Ocean.
2 million	First glacial episode in Northern Hemisphere.
1 million	First major interglacial.
100,000	Most recent glacial episode.
20,000–18,000	Last glacial maximum.
15,000–10,000	Melting of ice sheets.
10,000–recent	Present interglacial.

There was serious disagreement among geologists about the magical number of four ice ages. Strong evidence suggested the existence of several more. But many geologists, especially those in America, refused to accept this idea and simply chose to ignore it. Finally, in the mid-1950s, the Italian born climatologist Cesare Emiliana, working at the University of Chicago, produced unrefutable evidence for rapid and rhythmic successions of ice ages. His proof was obtained by analyzing the heavy oxygen content of fossil shells in cores taken from the bottom of the ocean. When the climate is cold, the heavier isotope of oxygen tends to remain when seawater evaporates and become concentrated in the shells of living organisms. Thus, by dating the fossils, Emiliana could recognize seven distinct ice ages that occurred over the last 700,000 years.

In the 1830s, the English geologist Charles Lyell proposed that changes in sea level were caused by fluctuations in the amount of ice on the land. He was also a strong supporter of the theory of uniformitarianism, in which the slow geological

Fig. 2-9. *Maximum glacial advances in North America from the youngest to the oldest: Wisconsin (solid line), Illinoian (crosses), Kansan (dashes), and Nebraskan (circles).*

processes operated much the same in the past as they do today. The ice ages, however, appeared to be an exception to the slow geological processes taking place on the Earth.

Lyell coined the term *Pleistocene* to mean a period of recent life based on the fossil record of modern organisms. Since widespread glaciation also occurred during this time, however, the Pleistocene epoch, which began about 2 million years ago, has since become synonymous with the ice ages. The Pleistocene epoch ended with the end of the last ice age. It was followed by the Holocene epoch, or age of truly modern life, which began about 10,000 years ago (TABLE 2-2).

The term *ice age* presently means a time when the ice sheets were at their maximum extent during the Pleistocene epoch. Over the past 1 million years, there appears to have been at least nine individual ice ages. The most recent ice age began about 100,000 years ago and peaked about 18,000 years ago, when one-third of the northern landmass was covered with ice. In North America, the ice was 2 miles or more thick in places and extended as far south as Oregon and New York. At the same time, ice also covered most of Great Britain and Northern Europe. Throughout the world, alpine glaciers existed on mountains that are presently ice free. Antarctica, part of Asia, Greenland, and western and southern South America were also covered with thick ice sheets.

Five percent of the Earth's water was locked up in glacial ice. Indeed, so much water was removed from the ocean that the sea level dropped as much as 400 feet lower than it is today. This exposed several land bridges and aided the migration of man and animals from one continent to another. Native Americans were able to cross over into North America from Asia via the Bering Strait some 30,000 years ago. The Earth was significantly colder during the ice age. The average yearly global temperature was roughly 10 degrees Fahrenheit lower than it is today.

TABLE 2-2. The Geologic Time Scale

ERA	PERIOD	EPOCH	AGE IN MILLIONS OF YEARS	FIRST LIFE FORMS
Cenozoic	Quaternary	Holocene	.01	
		Pleistocene	2	Man
		Pliocene	10	Mastodons
		Miocene	25	Saber-tooth tigers
	Tertiary	Oligocene	40	
		Eocene	60	Whales
		Paleocene	65	Horses
				Alligators
Mesozoic	Cretaceous		135	
				Birds
	Jurassic		180	Mammals
				Dinosaurs
	Triassic		230	
	Permian		280	Reptiles
	Carboniferous	Pennsylvanian	310	Trees
		Mississippian	345	Amphibians
				Insects
Paleozoic	Devonian		405	Sharks
	Silurian		425	Land plants
	Ordovician		500	Fish
	Cambrian		570	Sea Plants
				Shelled animals
Proterozoic			2500	Invertebrates
			3500	Earliest life
			3800	Oldest rocks
Archeozoic			4600	Meteorites

THE ICE CAPS

Greenland, which is the world's largest island (FIG. 2-10), was discovered by the Vikings in A.D. 982, following their discovery of Iceland a decade earlier (there is evidence that they continued on westward and landed in North America centuries before Columbus). For over four centuries, the Norse thrived on Greenland until advancing glaciers froze them out. An interesting theory suggests that the frozen island was called Greenland to entice people to settle there. However, the climate during this time was unusually warm so perhaps parts of the island were green after all. Then around 1450, the climate grew colder, glaciers crept toward the sea, and the Norse, unable to retreat, were never heard from again.

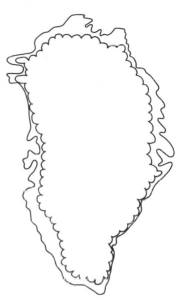

Fig. 2-10. The Greenland ice sheet.

Although its existence was predicted by Greek scholars some 2,000 years earlier, Antarctica (FIG. 2-11) was not discovered until about two centuries ago, and even then it was stumbled upon purely by accident. The British navigator James Cook discovered *terra incognita* (unknown land) in 1774, although heavy pack ice forced him to turn back before he could actually set eyes on the frozen continent. By 1820, the frigid waters around Antarctica were routinely visited by sealers, who slaughtered seals by the thousands for their oil and fine pelts.

The United States, Great Britain, France, and Russia sent expeditions into the south polar seas, during which the first official sightings of Antarctica were made. One of these expeditions was commanded by the Scottish explorer Sir James Clark Ross who attempted to find the South Magnetic Pole in 1839. He sailed his ships through 100 miles of pack ice on the Pacific side of the continent before finally emerging into open water, known today as the Ross Sea in his honor. With his way blocked by an immense wall of ice 200 feet high and 250 miles long, Ross gave up his quest to reach the South Magnetic Pole, which was about 300 miles inland from his position.

In 1902, the British explorer Commander Robert Scott attempted to reach the geographical South Pole from McMurdo Sound (FIG. 2-12), but was forced to turn back after traveling only about a third of the way to the Pole. In 1909, British explorer Ernest Shackleton, one of Scott's former team members, came within 112 miles of the Pole, but was forced to turn back due to low supplies and foul weather. However, one of Shackleton's teams successfully located the South Magnetic Pole, which in its own right was an important scientific achievement.

In 1911, Scott made a second attempt to reach the Pole, but this time he had competition from the Norwegian explorer Roald Amundsen. With his harden team of veteran explorers, Amundsen reached the Pole on December 15, a month ahead of Scott. He completed the 1,600-mile round trip in less than 100 days. Scott also reached the Pole, only to find that Amundsen had been there first. On the return trip,

(PHOTO BY W. B. HAMILTON, COURTESY OF USGS)

Fig. 2-11. The Antarctic ice sheet.

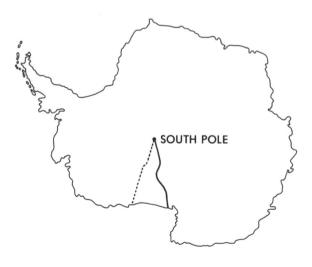

SOUTH POLE

Fig. 2-12. Routes taken by Scott (solid line) and Amundsen (dashed line) to the South Pole.

Scott and two of his companions were caught in a raging blizzard and froze to death just 13 miles from their supply depot. Amundsen later died in a tragic airplane crash while exploring the opposite end of the Earth.

There is still some controversy over who reached the North Pole first. The American Admiral Robert Peary has long been credited for reaching 90 degrees north on April 6, 1909. Starting out from Cape Columbia on Ellesmere Island, Peary reached

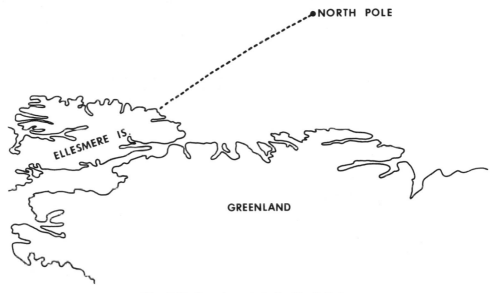

Fig. 2-13. Peary's route to the North Pole.

Camp Bartlet on April 1. He then made an eight-day dash to the North Pole and back during a spell of clear weather (FIG. 2-13). However, a study of his diary suggests that through compounding navigational errors, Peary might have missed the pole by as much as 30 to 60 miles.

Ancient Ice Ages

ICE ages dramatically affected life on Earth, practically from the very beginning. It is even argued that life itself might have sufficiently changed the climate to cause the ice. The gaia (pronounced guy-a) hypothesis, introduced by the English chemist-biologist James Lovelock in 1979, asserts that to a certain extent life can control its own environment in order to maintain optimum living conditions, much like the way the human body controls its temperature for maximum metabolic efficiency. Life clearly affects the composition of the oceans and atmosphere. Without life, the Earth's climate would be totally out of control.

When the first microscopic plants developed, they began to replace the carbon dioxide in the atmosphere with oxygen. The loss of this important greenhouse gas caused the climate to cool substantially, even though the Sun was getting progressively hotter. This cooling brought on the first known glacial epoch about 2 billion years ago. Another glacial episode that occurred some 260 million years ago might have been triggered by the spread of forests on the land as plants adapted to living and reproducing out of the sea. The Earth began to cool as the forests removed atmospheric carbon dioxide and converted it into organic matter that was buried in the sediments. This matter was then changed into coal.

The burial of carbon dioxide in the Earth's crust might have been the key to the onset of two other glacial epochs: one about 700 million years ago, which was perhaps the greatest of them all, and the most recent ice ages, which began about 2 million years ago. The positions of the continents had a tremendous influence on the initiation of the ice ages as well. Landmasses existing at higher latitudes allowed the buildup of glacial snow and ice. Global tectonics might also have triggered the ice ages through volcanic activity and seafloor spreading, which drew oxygen out of the oceans and atmosphere so that more organic carbon was preserved in the sediments and not returned to the atmosphere by living organisms.

THE DIM SUN

By the time the Earth cooled down enough to allow an ocean to form, the atmosphere had evolved into a mixture of water vapor, carbon dioxide, methane, unstable ammonia (which broke down into hydrogen and nitrogen), and traces of other gases. This was due to the outgassing of volcanoes and the degassing of meteorites and comets. The original atmosphere probably contained as much as 25 percent carbon dioxide. Presently, the atmosphere on Venus contains 98 percent carbon dioxide, which is responsible for a runaway greenhouse effect that keeps surface temperatures at 460 degrees Celsius (860 degrees Fahrenheit), hot enough to melt metals like lead and zinc.

Carbon dioxide is transparent to incoming sunlight, but absorbs and re-emits groundward outgoing infrared radiation. The early Earth was able to retain a good portion of the Sun's energy through the greenhouse effect (FIG. 3-1). This was fortunate because during the Earth's formative years the Sun's energy output was only about two-thirds of what it is today. If the early Earth had had our present atmosphere, the average global temperature would have been well below the freezing point of water. The entire ocean would have been a solid block of ice. However, because of the large quantity of atmospheric carbon dioxide, the Earth in its early years was much warmer than it is now.

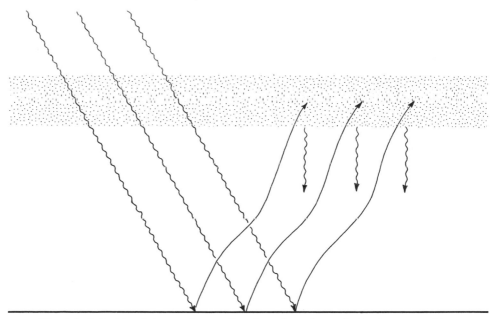

Fig. 3-1. Principle of the greenhouse effect: incoming solar radiation is converted into infrared radiation, which is absorbed by carbon dioxide and re-radiated back to Earth.

When the Sun became progressively hotter, carbon dioxide was taken out of the atmosphere and stored on the ocean floor and on the continents as carbonate rocks, which are comprised mostly of limestone. These early limestones were chemically precipitated and were not of biological origin. If carbon dioxide had been allowed to increase without these moderating factors, the Earth would have become hot enough to boil away its ocean, which may have been Venus' fate. Life would have had little chance for survival.

On the other hand, if the oceans and continents had continued to remove carbon dioxide from the atmosphere without resupplying it through volcanic activity, the Earth would still have been in trouble: the temperatures would have plummeted, and the planet would have been encased in ice. Luckily, carbon dioxide was replenished at nearly the same rate as it was removed, setting the global thermostat at a comfortable temperature for the existence of life.

As time progressed, green plants dominated the planet. They removed carbon dioxide, combined it with sunlight, and stored it in their tissues. The Earth probably cooled as a result of the evolution of surface plants. Nevertheless, it was still warmer then than it is today. In the great swamps of the Carboniferous period 350 to 280 million years ago, plant growth was highly prolific, probably due to abundant carbon dioxide in the atmosphere. Atmospheric oxygen then was also richer than at any other time.

The Carboniferous and Permian periods had the highest organic burial rates in the Earth's history. Plants formed thick layers of vegetative matter that was later buried and converted into coal. Extreme temperatures and pressures also cooked organic matter into oil and natural gas.

During the warmest part of the Earth's history, known as the Cretaceous period which existed 135 to 65 million years ago, land animals were plentiful and roamed practically from pole to pole. Volcanoes were particularly active and injected large amounts of carbon dioxide into the atmosphere. This significantly heated up the planet and might have been responsible in part for the giantism of the dinosaurs.

THE PROTEROZOIC ICE AGE

The first glacial epoch to be recorded in the geologic record took place about 2 billion years ago during the Early Proterozoic era. At this time, the landmass was composed of small odds and ends of granitic crust called cratons, which were assembled much like a giant jigsaw puzzle. Some of the original cratons formed within the first 1.5 billion years of the Earth's existence and totaled only about 10 percent of the present landmass. The Proterozoic era was also a period of transition, when atmospheric carbon dioxide was in the process of being replaced by oxygen generated by photosynthesis. The loss of this important greenhouse gas caused the planet to cool substantially and placed almost the entire landmass in the grips of a great ice age.

Organisms might have developed photosynthesis as early as 3.5 billion years ago, but the oxygen that they produced in this manner was removed by chemical processes that permanently buried it in the Earth's crust. While generating oxygen, simple plants also removed carbon dioxide from the environment and locked it up in the sediments. About 2 billion years ago, these oxygen repositories filled up and oxygen levels began to slowly increase in the ocean and atmosphere.

Along about this time, plate tectonics began to operate more extensively. The original landmass was probably located near one of the poles where ice sheets could easily grow. With the movement of the plates, carbonaceous sediments and the oceanic crust were thrust deep inside the Earth (FIG. 3-2). The growing continents also stored large quantities of carbon dioxide in thick deposits of carbonaceous rocks. The elimination of carbon dioxide, an important greenhouse gas responsible for maintaining the Earth's warm temperature, caused the Earth to substantially cool. Thus began the first ice age the world had ever known, although it was not the worst.

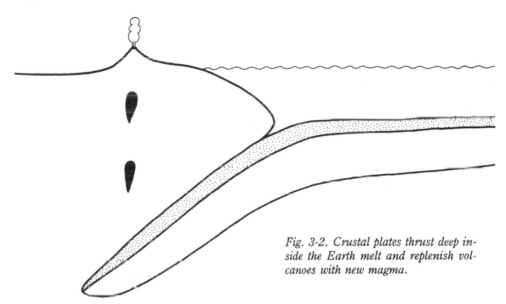

Fig. 3-2. Crustal plates thrust deep inside the Earth melt and replenish volcanoes with new magma.

THE LATE PRECAMBRIAN ICE AGE

The greatest glaciation the Earth has ever endured took place about 700 million years ago, when nearly half the planet was encased in ice. The climate was so cold that ice sheets and permafrost (permanently frozen ground) existed near equatorial latitudes. At this time, no plants grew on the barren landscape and only simple plants and animals lived in the sea.

The continents during Precambrian time were composed of various odds and ends of continental crust, or cratons. Four or five cratons were welded together to form what is now central Canada and the north-central United States. Continental collisions continued to add a large area of new crust to the growing proto North American continent. The better part of the underlying continental crust from Arizona to the Great Lakes to Alabama formed in one great surge of crustal generation around 1.8 billion years ago. This surge has been unequaled in North America since. It might have happened because plate tectonics and subsequent crustal generation operated at a much faster pace during Precambrian time than it does today.

The presence of large amounts of volcanic rock near the eastern edge of North America implies that it was once part of a supercontinent during middle Precambrian time. The central portion of the supercontinent was far removed from the cooling effects of the subducting plates, where the Earth's crust sunk into the mantle. As a result, the interior of the continent heated up and volcanoes erupted, while the warm, weakened crust broke apart. Toward the end of Precambrian time, about 700 million years ago, another supercontinent located near the equator possibly broke apart into four major continents, one of which apparently wandered into one of the polar regions and acquired a thick blanket of ice.

Thick sequences of Precambrian tillites are known to exist on every continent (FIG. 3-3). Tillites are a mixture of boulders, pebbles, and clay that was deposited by glacial ice and cemented into solid rock. In the Lake Superior region of North America, tillites are 600 feet thick in places and range from east to west for 1,000 miles. In northern Utah, tillites pile up to an impressive thickness of 12,000 feet. The various layers of sediment seem to indicate that there were a series of ice ages closely following each other. Similar tillites were found among Precambrian rocks in Norway, Greenland, China, India, southwest Africa, and Australia.

Fig. 3-3. Location of late Precambrian glacial deposits.

When the glacial epoch came to an end and the ice sheets retreated, life began to proliferate in the ocean with an intensity never seen before or since. Three times as many phyla, organisms that share the same general body plan, were alive then than are living today. As a result, many unique and bizarre creatures dominated the fossil record of that time (FIG. 3-4).

THE PALEOZOIC ICE AGE

Continental movements are thought to be responsible for a glacial epoch that occurred during the late Ordovician period about 440 million years ago. The study

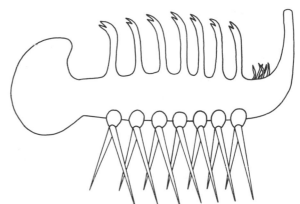

Fig. 3-4. Hallucigenia is one of the strangest animals preserved in the geological record. It walked on seven pairs of stilts and had seven tentacles, each with its own mouth.

of the magnetic orientation of rocks from many parts of the world indicates the position of the continents relative to the magnetic poles at various times in the Earth's history. One finding of these paleomagnetic studies was very curious, however, for it placed North Africa directly over the South Pole during the Ordovician period. Moreover, evidence for such widespread glaciation came from a surprising location—the middle of the Sahara desert.

Geologists exploring for oil in the Sahara stumbled upon a series of giant grooves that appeared to be cut into the underlying strata by glaciers. The scars were created when rocks, embedded at the base of glaciers, sawed into the landscape. Other collaborating evidence that showed that the Sahara desert had once been covered by thick sheets of ice included irratic boulders and sinuous sand deposits from glacial outwash streams called eskers (FIG. 3-5).

(PHOTO BY C. R. TUTTLE, CCURTESY OF USGS)

Fig. 3-5. Esker near Mallory, Oswego County, New York.

The glaciations of the late Ordovician period, the glacial epochs of the Permo-Carboniferous period around 290 million years ago, and the late Carboniferous period around 330 million years ago might have been influenced by about a 25 percent reduction in atmospheric carbon dioxide compared to its present value. Only recently have atmospheric scientists acquired enough information on global geochemical cycles to ascertain what might have caused the change in carbon dioxide concentration in the atmosphere.

New data from deep-sea cores indicate that carbon dioxide variations preceded changes in the size of more recent ice sheets. It is assumed that the earlier glacial epochs were similarly affected. The variation of carbon dioxide levels might not have been the sole cause of glaciation, however, but when combined with other processes, it could have been a strong influence.

The Late Paleozoic era was also a period of extensive mountain building, which allowed glaciers to be nurtured in the cold, thin air of higher elevations (FIG. 3-6). Glaciers might have formed and persisted on continents at low latitudes with high elevations because for every 1,000 feet of elevation, there is an equivalent increase of 300 miles of latitude. This means that the top of a 20,000-foot mountain located at the equator would be as frigid as the polar regions (FIG. 3-7).

Fig. 3-6. Alpine glacier on Jack Mountain, Washington.

Fig. 3-7. Alpine glaciation by latitude. Stippled areas represent the extent of forest growth.

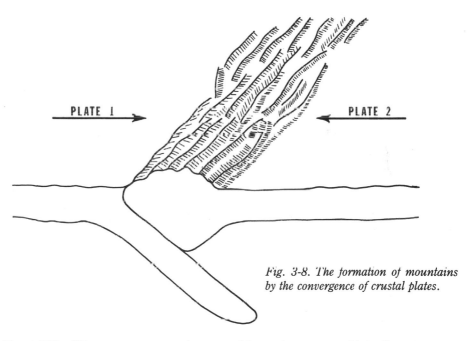

Fig. 3-8. The formation of mountains by the convergence of crustal plates.

About 280 million years ago, gondwana and laurasia converged into the crescent-shaped supercontinent of Pangaea, whose landmass extended almost from pole to pole. The continental collision crumpled the crust and pushed up huge masses of rocks into several mountain chains (FIG. 3-8). In addition to the folded mountain belts, volcanoes were also prevalent. Clouds of volcanic dust and aerosols from unusually long periods of volcanic activity might have blocked out the Sun and thereby lowered surface temperatures.

The Paleozoic Ice Age   **37**

As the continents rose higher, the ocean basins dropped. Changes to the shapes of the ocean basins greatly affected the course of ocean currents, which in turn had a profound effect on the climate. The continental margins became less extensive and narrower. The land that was once covered by great coal swamps dried out and the climate got colder.

During the breakup of the Pangaean landmass about 180 million years ago, when Gondwana separated from Laurasia and passed into the south polar region, glacial centers expanded in all directions. Land existing near the poles was often the cause of extended periods of glaciation. This is because land located at higher latitudes usually has a higher albedo and a lower heat capacity than the surrounding seas, and disrupts the poleward movement of oceanic heat.

Ice sheets covered large portions of east central South America, southern Africa, India, Australia, and Antarctica (FIG. 3-9). During the early part of the glacial epoch, the maximum glacial effects were felt in South America and southern Africa. Later, the chief glacial centers moved to Australia and Antarctica.

New evidence consisting of large, out-of-place boulders strewn across the desert in Australia suggests that ice existed there even during the warm Cretaceous period. At that time, the interior of Australia was filled by a large sea. Sediments settling on the floor of the basin were lithified into sandstone and shale and were later heaved upward high above sea level. Sitting in the middle of these sedimentary deposits are curious-looking boulders of exotic rock called drop stones, some measuring as much as 10 feet across. The boulders might have been rafted out to sea on slabs of glacial ice. When the ice melted the huge rocks dropped to the seafloor, where their impacts disturbed the underlying sediment layers.

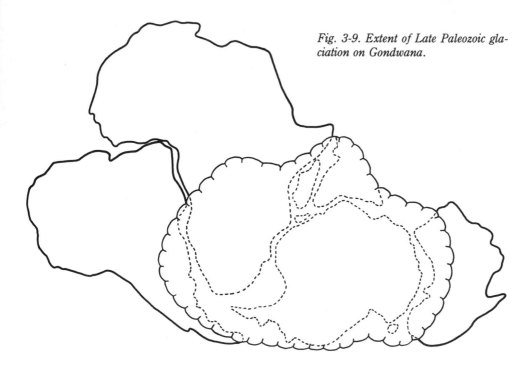

Fig. 3-9. Extent of Late Paleozoic glaciation on Gondwana.

Apparently, during the Middle Cretaceous period, Australia existed near the Antarctic Circle and was still attached to Antarctica. This provides clear evidence of the existence of Gondwana and the fact that the southern continents had wandered together over the South Pole. (See FIG. 3-10.)

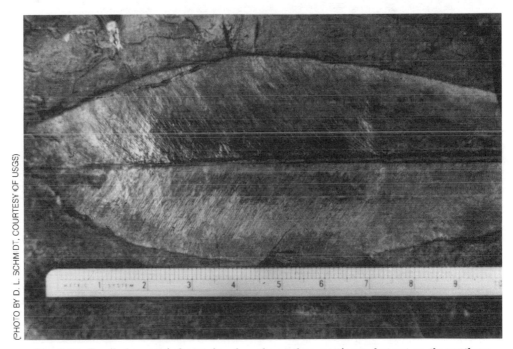

(PHOTO BY D. L. SCHMIDT, COURTESY OF USGS)

Fig. 3-10. Fossil glossopteris leaves found on the southern continents but not on the northern continents is evidence of the existence of Gondwana.

THE ICE-FREE CRETACEOUS PERIOD

During most of the Mesozoic era, about 240 to 65 million years ago, the climate warmed considerably. The steamy Cretaceous period, which began about 135 million years ago, had average global temperatures that were 20 degrees Fahrenheit warmer than they are today. There is no evidence of any permanent ice caps, such as those that today bury Antarctica and Greenland. The waters of the abyssal, which are now near freezing, were 25 degrees warmer. The temperature difference between the poles and the equator was only about 40 degrees, whereas today the difference is as much as 75 degrees.

The oceans of the Cretaceous period were interconnected in the equatorial regions by the Tethys and Central American seaways, providing a unique warm circumglobal current system. Coral reefs and other tropical life, for which warm water is essential, thrived as much as 1,000 miles closer to the poles than they do today. Polar forests extended into latitude 85 degrees north and south of the equator. The most remarkable example is a well preserved fossil forest on Alexander Island in Antarctica. Alligators and crocociles lived in latitudes (FIG. 3-11) as far north as Labrador, whereas today they are restricted to the warm Gulf Coastal region.

The drifting of the continents into warmer equatorial waters might have been a

Fig. 3-11. Where crocodiles once roamed, Uinta Mountains, Utah.

major contributing factor to the warm Cretaceous climate. The continents were flatter with lower mountain ranges and higher sea levels. The continents were also bunched together near the equator, which allowed the warm ocean currents to carry heat toward the poles. The high-latitude oceans were less reflective than land and therefore absorbed more heat, which further moderated the climate. The continents might have moved about much faster than they do today because of more vigorous plate motions. This produced greater volcanic activity.

Perhaps the greatest contribution to the warming of the Earth, came from this increased volcanic activity. The volcanoes added large amounts of carbon dioxide to the atmosphere, which substantially increased the greenhouse effect. The carbon dioxide also provided an abundant source of carbon for green plants. Intense and numerous lightning storms also fixed large amounts of nitrogen for plant growth. This contributed substantially to the diets to the dinosaurs and might have been one of the reasons why some grew so big.

During the Cretaceous period, great deposits of limestone and chalk were laid down in Europe and Asia. The period got its name from the Greek word creta meaning chalk. Seas invaded Asia, South America, Africa, Australia, and North America. Thick layers of sediment were deposited in the interior seaway of North America (FIG. 3-12) and are presently exposed as impressive sandstone cliffs in the western United States.

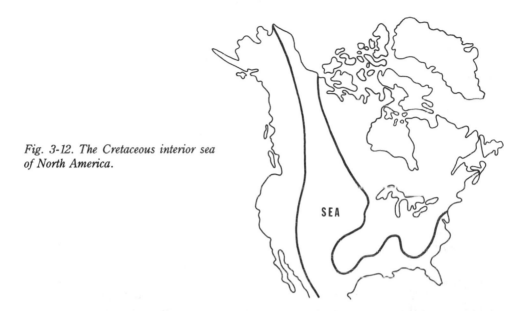

Fig. 3-12. The Cretaceous interior sea
of North America.

SEA

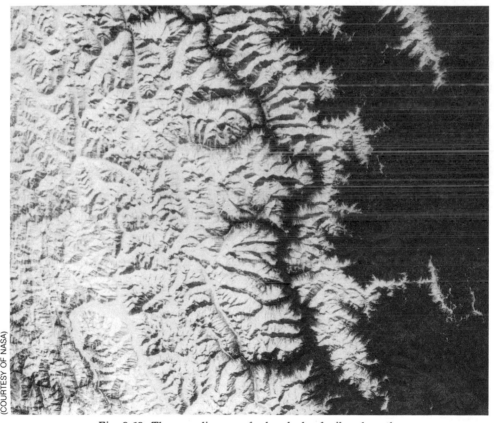

(COURTESY OF NASA)

Fig. 3-13. The snow line seen for hundreds of miles along the
southern face of the Himalayan Mountains in India.

The Ice-Free Cretaceous Period   41

The Appalachains, which were an imposing mountain range at the beginning of the Triassic period, 240 million years ago, were eroded down to stumps by the time of the Cretaceous period. The rim of the Pacific basin became extremely active geologically, and practically all the mountain ranges facing the Pacific and island arcs that run along its margins, such as the Aleutians, developed during this period.

Toward the end of the Cretaceous period, North America and Europe ceased to be connected, except for a land bridge created by Greenland to the north. The strait between Alaska and Asia narrowed and created the Arctic Ocean, which became nearly land-locked. The South Atlantic continued to widen, separating South America and Africa by over 1,500 miles of ocean. Africa moved north toward Eurasia. Antarctica, which was still joined to Australia, was left behind. As Africa and Eurasia came closer, they began to squeeze the Tethys Seaway until it almost totally dried up.

Meanwhile, India, which had split off of gondwana 100 million years ago, headed for southern Asia. The collision between the two landmasses uplifted the Himalayan Mountains (FIG. 3-13) and the broad Tibetan Plateau, which rises some 15,000 feet above sea level. As Antarctica and Australia moved eastward, a rift developed that eventually separated them. When Antarctica moved into the south polar region 40 million years ago, it acquired a permanent sheet of ice, which will remain until continental movements once again take the frozen continent into warmer seas.

4

Pleistocene Glaciation

WHEN the Cretaceous period ended 65 million years ago and the dinosaurs along with 70 percent of all other species became extinct, the Earth cooled down and for the first time developed two permanent polar ice caps. A single polar ice cap was a rare and short-lived event during the Earth's long history because for the most part, the Earth was warmer then than it is today and essentially ice free.

During the Pleistocene epoch, the most recent glacial epoch which spaned some 2 million years, numerous ice ages came and went almost like clockwork. The last ice age was a time when one-third of the Earth's land surface was covered with ice. Temperatures dropped by as much as 10 degrees Fahrenheit and sea levels lowered by as much as 400 feet.

THE COLD CENOZOIC ERA

The Cenozoic era is subdivided into the Tertiary period, which covers most of the era, and the Quaternary period, which covers the last 2 million years. Both terms were carried over from the old geologic time scale, in which the Primary and Secondary periods represented ancient Earth history. The unequal time distribution is in recognition of the unique, worldwide Late Cenozoic glacial epoch that began about 2 million years ago. Most geologist today prefer to use the terms Paleogene and Neogene, which more evenly divide the era.

The Cenozoic era was a time of constant change. All species including man, had to adapt to a wide range of living conditions. Changing climate patterns were brought on by the movement of the continents toward their present positions and intense tectonic activity, which built landforms and raised most of major mountain ranges of the world.

About 60 million years ago, a great rift developed between Greenland and Norway. Greenland also began to separate from North America. As recently as 4 million years ago, Greenland was largely ice-free, whereas today it is blanketed by an ice sheet up to 2 miles thick. Periodically, Alaska would connect with eastern Siberia and close off the Arctic basin from warm currents. This resulted in the accumulation of pack ice in the Arctic Ocean.

The Mid-Atlantic Ridge system (Fig. 4-1), which generates new ocean crust for the Atlantic basin, began to occupy its present position midway between North America and Eurasia about 16 million years ago. North and South America remained separated until the Panama isthmus was uplifted about 4 million years ago. Prior to this time, powerful currents flowed from the Atlantic into the Pacific and created a unique heat distribution system. South America was temporarily connected to Antarctica by a narrow, curved land bridge. Antarctica separated from Australia about 40 million years ago, drifted over the South Pole, and acquired a thick permanent ice sheet.

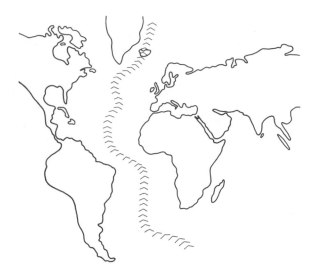

Fig. 4-1. The mid-Atlantic spreading ridge system that separated the New World from the Old World.

Intense mountain building took place during the Cenozoic era. There were major changes all along the Pacific Coast of North America. Volcanic activity was extensive and massive floods of lava covered Washington, Oregon, and Idaho. Basalts of the Columbia River Plateau covered 200,000 square miles and reached a thickness of 10,000 feet in places. The tall volcanoes of the Cascade Range (Fig. 4-2) from northern California to Canada erupted one after another. There was also extensive volcanism in the Colorado Plateau and Sierra Madre region.

The Rocky Mountains (Fig. 4-3) were raised during the Laramide mountain building episode, or orogeny, which began about 80 million years ago and ended some 40 million years ago. In the highest ramparts of the Rockies, glaciers developed when the Earth suddenly turned cold around 37 million years ago. A large part of western North America was also uplifted. The entire Rocky Mountain region was raised over

Fig. 4-2. The south side of Mount Adams in the Cascade Range, Washington.

3,000 feet due to increased buoyancy of the continental crust. This resulted from the subduction of vast amounts of oceanic crust beneath the west coast of North America during the Laramide orogeny.

Large numbers of parallel faults sliced through the basin and range province (or Great Basin) between the Sierra Nevadas and the Wasatch Mountains, beginning about 37 million years ago. This produced some 20 peculiar, north-south trending mountain ranges. The Great Basin area is a remnant of a broad belt of mountains and high plateaus that subsequently collapsed due to the pulling apart of the crust following the Laramide orogeny. Death Valley (FIG. 4-4), which is now 280 feet below sea level, was once several thousand feet higher. The area collapsed when the continental crust thinned out due to extensive block faulting in the region.

About 24 million years ago, a change in relative motions between the North American plate and the Pacific plate created the San Andreas fault system that runs

Fig. 4-3. Lake Sherburne Valley in the Rocky Mountains of Glacier National Park, Montana.

Fig. 4-4. Death Valley, California.

through southcentral California (Fig. 4-5). Presently, the Pacific plate is sliding toward the northwest, past the North American plate, at roughly 2 inches per year. The movement results in a strain that causes the San Andreas fault to rupture producing great earthquakes. Scientists believe that the southern end of the fault is due to rup-

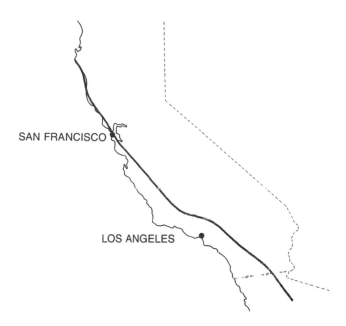

Fig. 4-5. The San Andreas fault of southern California.

SAN FRANCISCO

LOS ANGELES

ture at any time. At the northern end of the fault lies San Francisco which suffered severe earthquakes in April 1906 and October 1989. However, the latter one was not the great quake scientists had predicted, and was one of many tremors California has experienced over the last century and a half. As a result of other crustal movements, Baha California split off from North America and opened up the Gulf of California. Arabia also split from Africa. The rift formed the Red Sea and the Gulf of Aden.

The Tethys Seaway, which separated Eurasia from Africa, began to narrow about 50 million years ago as the two continents came together. The Tethys began to close entirely about 20 million years ago. This formed the Mediterranean Sea. Pinched off sections became the Caspian and Black Seas. Thick sediments that had been accumulating on the bottom of the Tethys for tens of millions of years were folded upward and formed great belts of mountain ranges on the northern and southern continental landmasses.

This episode of mountain building was called the Alpine orogeny. It ended about 30 million years ago and divided the Cenozoic era into the Paleogene and Neogene periods. It raised the Alps of northern Italy, the Pyrenees on the border between Spain and France, the Atlas Mountains of northwest Africa, and the Carpathians in east-central Europe.

The subcontinent of India, which had broken away from Africa in the early Cretaceous period, sped across the ancestral Indian Ocean and slammed into southern Asia about 40 million years ago. This collison uplifted the Himalyan Mountains and the wide Tibetan Plateau. The collision resulted in changes in plate motions, which apparently caused the Pacific plate to make a 45 degree left turn. A bend of the Pacific seamounts (FIG. 4-6), which are undersea volcanoes, bears testimony to this movement.

In South America, the Andes Mountains that run along the western edge of the continent continued to rise throughout much of the Cenozoic era. This was due to an increase in crustal buoyancy resulting from the subduction of the Nazca plate,

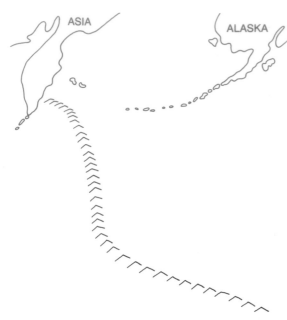

Fig. 4-6. The Emperor and Hawaiian seamounts represent motions on the Pacific plate over a volcanic hot spot deep within the mantle.

which is subducted beneath the South American plate. By the time all the continents had wandered to their present positions, and all the mountain ranges had risen to their present heights, the world was ripe for the development of the Pleistocene ice sheets.

QUATERNARY GLACIATION

The Pleistocene epoch, which began about 2.4 million years ago, witnessed a progression of ice ages. Each ice age was followed by a short interglacial age similar to the one in which we are presently living. The last ice age began about 100,000 years ago, intensified 75,000 years ago, peaked about 18,000 years ago, and retreated about 10,000 years ago. It appears that the ice took longer to build to its fullest extent than it took to recede to its present position at the poles. In what is only a geological moment in time, the ice sheets suddenly collapsed and rapidly disappeared.

The most recent glaciation is perhaps the best studied of all the ice ages. This is because each preceding ice age was mostly erased by the previous one when the ice bulldozed its way across the land. In some areas, the ice stripped off entire layers of sediments and left behind bare bedrock. In other areas, older deposits were buried under thick deposits of glacial till, which formed elongated hills called drumlins (FIG. 4-7).

During the last ice age, about 5 percent of the planet's water was locked up in glacial ice. This resulted in an appreciably lower sea level, which in turn expanded the land area by 8 percent (FIG. 4-8). Coral, which lives only in warm surface waters, fluctuated in height in response to the changing sea levels. As the buildup of glacial ice lowered the level of the sea, the coral reef eroded to the new sea level. When the glaciers melted and the sea level rose, new coral began to grow on top of the old, forming a terrace of coral growth. Alternating changes in that sea level corresponded

Fig. 4-7. Drumlins in the Province of Saskatchewan, Canada.

Fig. 4-8. Extended shoreline during the
height of the last ice age.

to the waxing and waning of the glaciers and produced a staircaselike structure of coral. Such coral terraces in the tropics, when dated, provide fairly accurate ages for glacial events.

Additional evidence of the rapid and rhythmic successions of glaciation came from analyzing the ratio of oxygen isotopes in the fossil shells of tiny organisms in rock cores taken from the ocean floor. The ratio of oxygen-18 (0–18) to oxygen-16 (0–16) is an indication of the ocean's past temperatures and provides a means of accurately dating the ice ages. Because 0–16 is the lighter of the two isotopes water molecules made of 0–16 evaporate more easily than those made of 0–18. This is especially true in colder climates when a higher concentration of 0–18 is left behind in the ocean.

Both oxygen isotopes are incorporated in the shells of marine organisms that were alive at that time. When fossils of these tiny creatures are anaylzed for their 0–16 and 0–18 ratios and dated, an accurate calendar for the ice ages is produced. Accordingly, there appears to have been nine full ice ages in the past million years, or one roughly every 100,000 years.

Ice cores taken from bore holes drilled in the glacial ice in Antarctica and Greenland have also been analyzed for their 0–16 and 0–18 ratios. During times of glaciation, the amount of 0–16 is found in greater abundance than 0–18, which is left behind in the ocean in colder climates. In addition, the ice contains airborne particles, including dust, volcanic ash, and sea salt, that fell on the glacier along with fresh snow. During periods of glaciation, the climate was both colder and dryer. This expanded the area of arid zones. Strong winds that blew across the deserts raised huge dust clouds (FIG. 4-9). The fallout from these clouds ended up on top of the glaciers.

(PHOTO BY CORNELIUS KEYES, COURTESY OF NOAA)

Fig. 4-9. A dust storm rises over Phoenix, Arizona, on Labor Day 1972.

Fig. 4-10. Extent of the glaciers during the last ice age.

During times of intense volcanic activity, large amounts of volcanic ash were injected into the atmosphere along with sulfur gases, which precipitated as acid rain. The acid improved the electrical conductivity of the ice core, which enabled electrodes attached to the ice to detect periods of volcanism. These measurements agree with the historical record of major volcanic eruptions.

At the height of the last ice age (FIG. 4-10), ice as thick as 10,000 feet covered Canada, Greenland, and northern Europe. In North America, there were two main glacial centers. The largest glacier, called the Laurentide ice sheet, covered an area of 5 million square miles. From the Hudson Bay it reached northward into the Arctic Ocean and also buried all of eastern Canada, New England, and much of the rest of the northern half of the midwestern United States. A smaller glacier, called the Cordilleran ice sheet, originated in the Canadian Rockies and engulfed western Canada, parts of Alaska, and small portions of the northwestern United States.

There were also two major glacial centers in Europe. The largest glacier, called the Finoscandinavian ice sheet, radiated out from northern Scandinavia. It covered Great Britain as far south as London, as well as large parts of northern Germany, Poland and European Russia. A smaller glacier, called the Alpine ice sheet, was centered in the Swiss Alps and covered parts of Austria, Italy, France, and southern Germany. In Asia, the ice sheets occupied the Himalayas and parts of Siberia.

In the Southern Hemisphere, small ice sheets expanded in the mountains of Australia, New Zealand, and the Andes of South America. Just about everywhere, alpine glaciers existed on mountains that are presently ice free. Only Antarctica had a major ice sheet, which grew about 10 percent larger than its present size. The excess ice had nowhere else to go except into the ocean, where it broke apart to form icebergs

(FIG. 4-11). During the peak of the last ice age, icebergs covered half the area of the oceans. Their high albedo reflected a great deal of sunlight out into space, which cooled the Earth and allowed the glaciers to grow.

The surface temperature, when averaged over the entire globe and over the seasons, was about 10 degrees Fahrenheit lower than it is today. The cold weather and advancing ice forced animals and humans to migrate to the warmer southern lands. Ahead of the slowly advancing ice sheets, which might have increased at most only a few hundred feet per year, lush deciduous forests gave way to evergreen forests. These in turn gave way to grassland and finally to barren tundra and rugged periglacial regions that existed at the margins of the ice sheets (FIG. 4-12).

(COURTESY OF U.S. NAVY)

Fig. 4-11. A huge iceberg off the coast of Graham Land on the Antarctic Peninsula.

(PHOTO BY R. B. COLTON, COURTESY OF USGS)

Fig. 4-12. Antonelli Glacier showing rugged periglacial area,
including recessional and other moraines.

Less water evaporated from the oceans because of the lowered temperatures, which reduced the average amount of precipitation. Since very little melting took place in the cooler summers, only a small amount of snowfall was needed to sustain the ice sheets. The lower precipitation levels also increased the size of the deserts in many parts of the world. Desert winds, which were much more blustery than they are today, produced gigantic dust storms. High levels of dust in the atmosphere blocked out sunlight, thus shading the Earth and keeping it cool. Large numbers of icebergs calved off glaciers entering the sea. Like ice cubes in a cold drink, they kept the ocean surface temperatures cool (FIG. 4-13).

Fig. 4-13. Areas of August sea-surface cooling during the last ice age.

When sea ice extended its reach northward and shaded the algae below, production of diatoms, one-celled algae with a shell made of silica (FIG. 4-14), sharply decreased in the surface waters of the Antarctic. Without sunlight for photosynthesis, the diatoms simply disappeared approximately 2.4 million years ago. This is also the agreed upon date for the initiation of Pleistocene glaciation in the Northern Hemisphere.

THE ICE AGE PEOPLES

Our present glacial epoch takes place in the Quaternary period, which consists of the Pelistocene and Holocene epochs. The Pleistocene epoch has become synonymous with the most recent ice ages. What is noteworthy about this particular glacial epoch is that humans first appeared on the scene just about the time the glaciers started to spread across the northern continents some 2 million years ago, so that ice ages span the whole of man's existence.

During the last ice age some 35,000 to 10,000 years ago, a period anthropologists call the Upper Paleolithic period, a species known as Cro-Magnon, developed a rich culture. These people were named for the Cro-Magnon cave in France where our

(PHOTO BY G. W. ANDREWS, COURTESY OF USGS)

Fig. 4-14. Fossil diatoms from the Choptank formation in Calvert County, Maryland.

human ancestors once lived. This was a time that witnessed an explosion of human progress and creativity and served as a milestone in human culture. Humans made more technological and artistic advancements in these 25,000 years than in the entire 2 million years since man first made stone tools.

In a quantum leap forward, humans invented language, art, and music. They also laid down the foundations of laws, trade, class distinctions, and fashion. These late ice age peoples made elaborate and beautiful drawings and carvings of animals, especially the ones that they had a great respect for and did not eat such as horses and cave bears. They invented sewing needles made of bone to tailor cold-weather clothes, which aided them in colonizing the colder regions of Europe. The Cro-Magnon buried their dead and adorned the graves with the personal belongings of their fallen kin, along with flowers and abex horns. They also developed more efficient hunting weapons. As the ice retreated, they followed the game animals northward and settled in northern Europe, Asia, and North America.

Surprisingly, there seems to have been little tribal warfare, so it is doubtful that the Cro-Magnon murdered the Neanderthals, their stocky primitive cousins. The Neanderthals roamed the cold climes of Europe throughout most of the ice age and were the first true ice age peoples because they survived despite the harsh cold. Perhaps their bulky bodies and large muscles enabled them to produce more body heat, which allowed the Neanderthals to endure the rigors of the ice age. Because they were very shy the Neanderthals probably fled from the advancing Cro-Magnon and eventually wandered into desolate regions and vanished. The Cro-Magnon then became the only human species.

The Cro-Magnon, which were like modern humans, might have originated in Africa as early as 90,000 years ago. Why they took some 50,000 years to penetrate farther into the Old World still remains a mystery. These people shared many of our current physical attributes. Their cranial capacity was as large as ours, and their brain case proportions were modern in structure. The skull was short, high and rounded, without the large brow ridges of Neanderthal skulls. The face was robust and flat like that of modern Europeans, and the lower jaw ended in a definite chin. The skeleton was slender and long-limbed in contrast to the stocky Neanderthals.

Sometime during the last glacial period, between 45,000 and 35,000 years ago, Cro-Magnon advanced into Europe and Asia, probably during a warm interlude when the ice age climate was not too severe. A rich stock of large animals might have attracted these people to desolate, frozen wastelands. During the last glacial period; the European geography was largely composed of grasslands instead of tundra. On this range roamed reindeer, wooly mammoths (FIG. 4-15), and other species specially

Fig. 4-15. The wooley mammoth became extinct about 11,000 years ago, along with other large mammals, toward the end of the last ice age.

adapted to the cold. The Cro-Magnon hunted these animals extensively and carved up the carcasses with finely honed blades. Deadly spears, often engraved with animals, became effective weapons for hunting big game and might have led to the extinction of certain species that were easy prey to the advanced hunting skills of humans.

Ice age peoples probably lived much like present-day Eskimos and Lapps. They fished the rivers, possibly using small boats, and hunted reindeer and other animals. Due to a lack of wood, ice age hunters on the central Russian plain built homes out of mammoth bones and tusks that were covered with animal hides. They burned bones and animal fat for light and heat in the frigid climate.

By crossing a land bridge between Siberia and Alaska, ice age peoples were able to populate the Americas as early as 32,000 years ago (FIG. 4-16). The massive continental glaciers locked up great quantities of water, which substantially lowered the sea level and exposed parts of the shallow sea floor. So much water was missing from the ocean that the North Sea, the Baltic Sea, and the Bering Strait did not exist. From North America, these early peoples crossed over the Panama isthmus into South America and roamed as far as southern Chile.

The peopling of the Pacific took place at a time when the sea level had dropped several hundred feet. Entire island chains were linked together, allowing a human migration of immense proportions. Humans occupied the distant continent of Australia as far back as 40,000 years ago. By island hopping from southeast Asia, polynesian people managed to settle the great ocean reaches. It is even contended that they went as far east as South America.

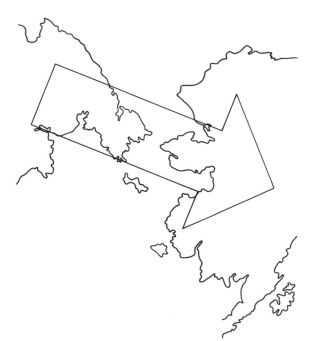

Fig. 4-16. The advance of ice age peoples from Asia into North America via the Bering Strait.

THE GIANT MAMMALS

Adaptation to the cold climate allowed certain species of mammals to thrive in the northern lands that were not covered by glaciers. Giant mammals like mammoths, mastodons, saber-tooth cats, and 20-foot-tall ground sloths inhabited many parts of the world. Their large bodies retained heat better than smaller ones, but required a great deal of food. The vegetation the giant mammals ate was probably coarse and of poor nutritional value, which meant that they needed vast quantities of food. This required a large stomach for the long fermentation process, which in turn necessitated a large body.

As the glaciers started to retreat about 16,000 years ago, there was a readjustment in the global environment. The cool but equable climate of the ice age gave way to the warmer, more seasonal climate of the Holocene epoch. The rapid environmental switch from glacial to interglacial caused the forests to shrink and the grasslands to expand. This might have disrupted the food chain of several of the larger mammals. Deprived of their nutritional resources, they simply disappeared.

All together, some 35 classes of mammals, including 10 classes of birds, became extinct in North America almost at the same time, between 12,000 and 10,000 years ago, with most occurring about 11,000 years ago. The vast majority of mammals affected were plant-eaters that weighed over 100 pounds. Some weighed as much as a ton or more. Unlike earlier episodes of mass extinction, such as the disappearance of the dinosaurs and 70 percent of all other species 65 million years ago, this one did not significantly affect small mammals, amphibians, reptiles, or marine invertebrates.

To make the mystery even more intriguing, archaeologists found that from 11,500 to 11,000 years ago, many parts of North America were occupied by ice age peoples. Their spear points have been found among the remains of giant mammals, including mammoths, mastodons, tapirs, native horses, and camels (FIG. 4-17). When these peo-

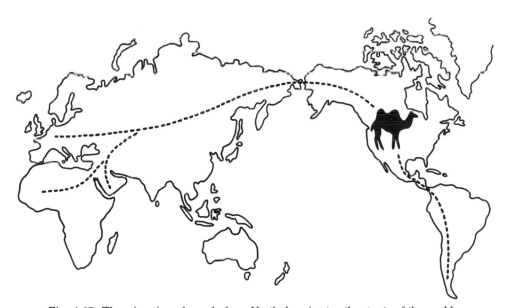

Fig. 4-17. The migration of camels from North America to other parts of the world.

ple crossed the land bridge over the Bering Strait and moved through the ice-free corridor east of the Canadian Rockies, they entered a happy hunting ground populated with upwards of 100 million large mammals, many species of which had been decimated in Europe and Asia. Extinction was also extensive in Australia, possibly caused by the ancestors of the aborigines.

As the climate warmed the sea level rose and the water levels over much of North America fell. The large mammals probably congregated at the few remaining water holes, where they became more vulnerable to hunting. With plentiful prey and exposure to few, if any, new diseases, the human population grew. People spread southward, killing entire populations of many big-game animals in their wake.

5

The Holocene
Interglacial

ODAY, we live in a glacial epoch that has lasted over 2 million years, with a succession of ice ages every 100,000 years or so. The ice ages persisted for some length of time, then suddenly the continental glaciers melted, and the climate warmed. Most previous interglacial periods lasted only 8,000 to 12,000 years. The fact that the last ice age ended some 10,000 years ago indicates that the present warm period probably has no more than a couple of thousand years to go before the beginning of a new ice age. The timing mostly depends on how humans alter the climate.

The beginning of a new ice age might be very gradual. Over the short run it may hardly be noticed. It might take upwards of tens of thousands of years before the ice sheets reach their maximum extent. Then the ice sheets might rapidly recede until most of the northern lands are once again free of ice. During the last ice age, it took the ice sheets about 70,000 years to reach their maximum size. It took only 10,000 years for them to shrink to a comparably small amount of ice in the Arctic (FIG. 5-1).

THE RETREATING ICE

One of the most dramatic climate changes in the history of the planet occurred during our present interglacial known as the Holocene epoch. It began about 10,000 years ago and is commensurate with the rise of modern man. The collapse of the giant ice sheets and the subsequent warm climate left behind many puzzles. One such mystery was the discovery of hippopotamus and crocodile bones in the middle of African deserts.

When the ice started to shrink, Africa and Arabia, which were tropical regions at the time, began to dry out. This resulted in an expansion of the arid regions be-

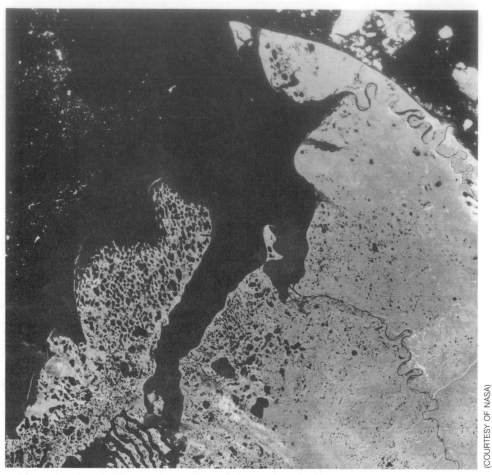

Fig. 5-1. Landsat 1 view of arctic tundra in Northwest Territories, Canada, East of Mackenzie Bay with drift ice north of Cape Bathurst.

tween 14,000 and 12,500 years ago during a period of rapid warming. A wet period that followed from 12,000 to 6,000 years ago caused some of today's African deserts to bloom and be dotted with several large lakes. Lake Chad, which lies on the border of the Sahara desert, was the largest lake of all and apparently covered an area 10 times its present size. Lakes in other parts of the world were similarly affected. Utah's Great Salt Lake occupied the adjacent salt flats and expanded to several times its current size (FIG. 5-2). African swamps, long since vanished, once harbored large populations of hippopotamuses and crocodiles, whose bones now bake in the hot desert sun.

Some of the most baffling questions concerning the Holocene climate are answered by analyzing pollen grains recovered from ancient bogs and lake bed sediments as well as other paleoclimate indicators such as plant fossils found in pack rat middens. Ancient plant and animal remains are compared to present-day relationships between species distribution and climate. For example, a combination of spruce

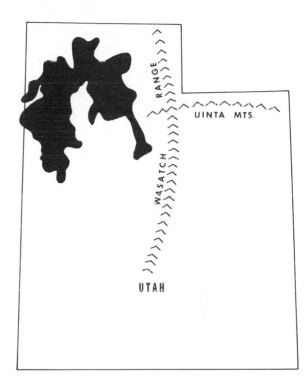

Fig. 5-2. The ancestral Great Salt Lake.

and sedge fossil pollen is an indicator of a cold, dry climate. Fossil pollen of leafy herbs typical of the prairies of the midwestern United States suggest a warm, dry climate.

Another important continental climate indicator is lake level fluctuations. Lakes act like natural rain gauges. Ancient rainfall amounts are implied by studying the positions of past shorelines and such depth indicators as the mineral, floral, and faunal composition of lake-bed sediments. Lake-level records are particularly valuable for the arid regions of Africa, Australia, and the western United States because pollen records are sparse in these regions.

Another paleoclimate indicator is the size of sand grains deposited over the past quarter of a million years in Lake Biwa, Japan, one of the oldest lakes in the world. The rise and fall of the sediment size kept pace with the advance and retreat of the ice sheets. The grain sizes reflect past erosion rates, which varied with such climate changes as rainfall, temperature, and wind speed. High precipitation rates increased erosion and therefore the amount of coarse sand grains that were carried into the lake.

After some 90,000 years of gradually accumulating snow and ice up to 2 miles thick in parts of North America and Eurasia, the glaciers melted away in only a few thousand years, retreating upwards of 600 yards per year. The rapid deglaciation might have been driven by forces other than the warming of the climate. The ultimate cause for bringing the last ice age to an abrupt end might have been due to changes in the planet's orbit and the tilt of its axis. The axis of rotation was more steeply tilted toward the Sun 9,000 years ago than it is today. The Earth came closest to the Sun in late July instead of early January as it does now. This made the seasonal change of distances between the Sun and Earth greater (FIG. 5-3).

In the Northern Hemisphere, which contained most of the glacial ice, the Earth received 7 percent more solar radiation in July, making the summers hotter than they are today. In addition, the high latitudes received the maximum amount of insolation (input of solar radiation) because the rotation of the axis was more inclined toward the Sun. The net effect of all this was that much of the Northern Hemisphere received 15 percent more sunlight during the summer than it did during the winter. This is about twice the difference in sunlight received today. The wide difference in sunlight made the contrast between seasons much greater. After 9,000 years, the seasonal climate extremes gradually decreased to more modern values.

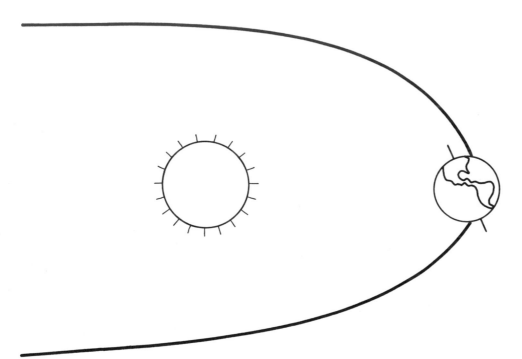

Fig. 5-3. The Earth 9,000 years ago came closest to the Sun in late July and the tilt of its spin axis was greater, allowing more sunlight to reach the northern latitudes.

At least one-third of the ice melted between 16,000 and 13,000 years ago, when average global temperatures increased by about 10 degrees Fahrenheit to nearly what they are today. As the North American ice sheet began its retreat, its meltwater flowed down the Mississippi River into the Gulf of Mexico. After the ice sheet retreated beyond the Great Lakes, however, the meltwater flowed down the Lawrence River and the cold waters entered the North Atlantic. It was also about this time that the Niagara River Falls (FIG. 5-4) began cutting its gorge. The gorge has expanded over 5 miles northward since the ice sheet retreated.

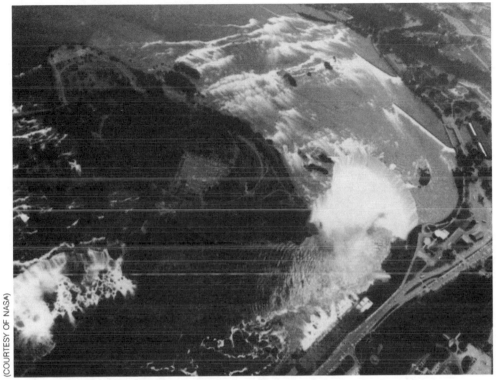

Fig. 5-4. Niagara Falls has receded up its gorge over 5 miles since the ice sheets retreated.

The rapid melting of the glaciers culminated in the extinction of microscopic organisms called foraminifera (FIG. 5-5). A torrent of meltwater and icebergs spilled into the North Atlantic and put a cold, fresh-water lid on top of the ocean. This significantly changed the salinity of the seawater. The cold waters also blocked poleward flowing warm currents from the tropics and caused the temperature on the land to fall to near ice age levels. As a result, the ice sheets appeared to pause in mid stride between 11,000 and 10,000 years ago. This period is called the Younger Dryas and is named after a flower that lives in the Arctic.

After that, the warm currents were re-established and the warming trend stayed for good. This prompted a second episode of melting, which led to the present volume of ice by about 6,000 years ago. This in turn led to the beginning of the so-called Climatic Optimum, a period of unusually warm and wet conditions that lasted for 2,000 years.

The long wet spell following the retreat of the glaciers is believed to have been caused by the strengthening of the monsoons, which carry moisture-laden sea breezes inland over Africa, India, and Southeast Asia. The interior of the continents 9,000 years ago warmed more in the summer, which strengthened the monsoon winds and increased rainfall. In North America, the glaciers radically changed atmospheric circulation and likewise the storm patterns over the continent. High pressure centers over the Laurentide ice sheet brought strong easterly winds across the southern flank

Fig. 5-5. Foraminifera became extinct at the end of the last ice age.

of the glacier and strengthened the jet stream aloft (FIG. 5-6). This made it possible for the Mojave and nearby deserts of the southwestern United States to receive enough rainfall to sustain woodlands after the ice sheets began their retreat.

From about 9,000 to 6,000 years ago, when the glaciers over North America shrank to a nearly insignificant chunk of ice in Labrador, precipitation over much of the midwestern United States dropped by as much as 25 percent. At the same time, the mean July temperatures rose by as much as 4 degrees Fahrenheit. It also appears that the post glacial eastern and southeastern United States was not significantly warmer 6,000 years ago than those areas are now, probably due to the temperature moderating effect of the Atlantic Ocean.

Then between 6,000 and 4,000 years ago, during the Climatic Optimum, many regions of the world warmed by an average of 5 degrees. The melting ice caps released a torrent of flood water into the sea and raised sea levels 300 feet above where they were at the beginning of the Holocene epoch.

Paleoclimate studies, which analyzed marine microfossils in the cores of seafloor sediments for their carbon dioxide levels, suggest that the previous warm interglacial (called the Sangamon), which ended 127,000 years ago, might have been even warmer than the present one. It appears that there was a higher concentration of atmospheric carbon dioxide during the Sangamon interglacial, which led to a greater amount of greenhouse warming. Ancient river beds that carried alluvium to the sea were also used as a source of information on changing climatic conditions of the past. These studies and others indicated that glacial periods existed many times longer than interglacial periods. The previous four interglacials lasted between 8,000 and 12,000 years. The fact that the present interglacial has existed for 8,000 to 10,000 years means that perhaps it has just about run its course.

Fig. 5-6. Typical flow of the jet stream.

Although the preceding interglacial might have been warmer than the present one, the warm climate was powerless in halting the last ice age. Increased greenhouse warming due to today's industrialization, destruction of the forests, and extension of agriculture might not in the long run save the human race from advancing ice sheets.

THE BIRTH OF AGRICULTURE

About 15,000 years ago, when the continental glaciers began to retreat, primitive peoples stumbled upon the fertile Levant region, which borders the Eastern Mediterranean Sea, and the area known as the Fertile Crescent, which stretches from the southeast coast of the Mediterranean Sea to the Persian Gulf. In these regions, abundant stands of wild wheat and barley grew in thickets on the uplands.

The stone-age peoples gathered the wild plants and used primitive stone grinders to process the cereals. The stability of this food supply encouraged people to build permanent settlements. They devised a number of tools to harvest the crop and invented pottery in which to store and cook it. They began herding wildlife instead of hunting them. This led to a new system of animal husbandry. Wild dogs were also tamed and trained to manage several chores, including rounding up herds and warning of danger.

The Neolithic age, or new stone age, was ushered in when the majority of the glaciers were gone, about 10,000 years ago. It was also the beginning of a food-producing revolution. Man slowly learned to control the resources at his disposal. Instead of merely gathering wild foods, he collected plants and attempted to control and nurture them. Sheep, goats, pigs, and cattle were domesticated, providing a ready supply of wool, milk, and meat. The animals also provided an alternative source of

Fig. 5-7. The spread of civilization following the retreat of the glaciers in Europe.

food if the harvest failed. When resources were exhausted, as they often were, the community collapsed and people were forced to move to a new location.

Even at this early stage, farming was so productive it could support 10 times more people in a given area of land than foraging. It is not surprising that agriculture not only supported but encouraged population growth. Because farming was so labor intensive large families were needed to till and harvest the land. Increased population, however, required people to intensify food production, often with disastrous consequences to the land.

Agriculture brought with it sedentism. With the building of villages and land ownership, people became territorial. Land and permanent houses had to be guarded and tended with care by the whole village so that they could be passed on to the next generation. Overpopulation, however, forced people to migrate out of the region. Farmers roamed across Europe, following the tracks of the hunter-gatherers, who themselves followed the retreating glaciers northward (FIG. 5-7).

Beginning about 6,000 B.C., agriculture spread from Greece through the Balkans. By about 3,000 B.C., it had reached northern Europe and Great Britain. Europe at this time was largely forested, and huge tracks of land were cleared for planting and cattle grazing. It is suspected that the felling of so many trees might have altered the climate, which in this region was cold and wet. The expansion of farming into the areas with the most easily cultivable soils again brought on overpopulation. This forced people to move to the less desirable regions such as the plains of Mesopotamia in southwest Asia, where the problem was not the lack of fertile soil, but the need for water.

Fig. 5-8. Pueblo Bonito ruins in Chaco Canyon, New Mexico.

The Neolithic age was a time of unprecedented human advancement. The dawn of agriculture at the beginning of the Neolithic age has been described as the greatest achievement and the worst mistake in the history of the human race. Although it freed societies from the constant search for food and gave people more time to contemplate other aspects of life, it also made people territorial by allowing them to settle in one place (FIG. 5-8).

HUMAN INDUSTRY

The boat was first introduced during the Neolithic age to help early mariners travel around the rim of the Mediterranean Sea. The first sailors did not want to lose sight of land for fear of wandering off into oblivion or getting caught in a raging storm. When the compass was invented around 2,000 B.C., sailors could chart their courses without fear of getting lost. The first sailing ships were introduced around 3,000 B.C., although they were rigged with square sails and could only travel in the direction of the wind. After reaching their destination, sailors had to wait for favorable winds to take them home.

People landed on the island of Cyprus in the eastern Mediterranean (FIG. 5-9) more than 10,000 years ago. The human presence on this island coincides with the disappearance of the pygmy hippopotamus, an animal about the size of a small pig that once roamed Cyprus and other nearby islands. One site on the island contained a large number of skull bones from these animals along with rock flakes that were apparently made by humans. It appears that this site might have been a refuse pile

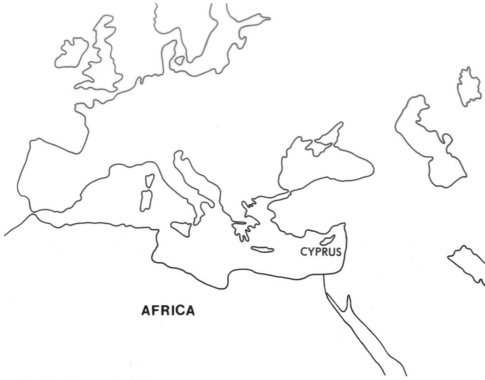

Fig. 5-9. The spread of civilization first took place around the rim of the Mediterranean Sea. Note the location of Cyprus.

and supports the possibility that humans had something to do with the extinction of the Cyprus hippopotamus.

The manufacturing of pottery and plaster around 7,000 B.C. was the first Neolithic industry. Plaster artifacts found in the Near East include flooring material, containers, sculptures, and ornamental beads. Plaster was probably invented as early as 12,000 B.C., long before pottery. With the introduction of agriculture, more durable storage vessels for agricultural goods were needed, which necessitated the invention of pottery.

Humans discovered fire early in the Pleistocene epoch. Without fire people might not have survived in the northern latitudes during the ice age. It is not clear, however, whether they could ignite a fire at this time. Most likely, they utilized fires naturally set by lightning strikes. For thousands of years, humans used fire to cook their meals, keep warm in winter, and hunt game by setting brush fires that frightened the animals into traps or forced them off cliffs.

About 6,000 years ago, people began to use fire to bake pottery and forge bronze for tools and weapons. The Bronze Age gave way to the Iron Age, about 1,000 B.C., which in turn gave way to the Middle Ages from about A.D. 500 to 1,500. Man's use of fire was considered rudimentary until the introduction of the steam engine in the late 1700s. The steam engine mated coal with water and ushered in the Industrial Revolution.

THE LITTLE ICE AGE

From about the time Columbus made his three voyages to America in the late fifteenth century until about the middle of the nineteenth century, the world was in the grips of a little ice age. Average yearly temperatures fell as much as 2 degrees Fahrenheit. Glaciers that had been retreating steadily since the end of the last ice age suddenly began to advance. The Vikings, who had settled on Greenland about A.D. 980 and successfully raised cattle and crops, were frozen out by the expanding ice sheets. In the northlands of Europe, creeping glaciers chased people out of what were once lush valleys. In the late eighteenth century, when American colonists fought against Britain during the Revolutionary War, the severe winters threatened the colonial army more than the British.

When the war was over, Benjamin Franklin went to Paris, France, to become the first diplomatic envoy from the newly created United States of America. During the summer, Franklin noticed a constant dry fog all over Europe and North America that substantially reduced the temperature. The winter of 1783-84 was one of the most severe winters on record. Franklin attributed the strange weather phenomenon to volcanic ash in the atmosphere from the eruption of the Icelandic volcano Laki, which killed 10,000 people and 200,000 livestock. The volcanic ash spread across the Northern Hemisphere during the summer of 1783 and blocked the light from the Sun, which shaded and cooled the Earth.

It appears that Icelandic eruptions (FIG. 5-10) were more frequent during warm

(COURTESY OF U.S. NAVY)

Fig. 5-10. Birth of a new volcano near Geirfuglasker Island, Iceland.

interglacials when Iceland was largely free of ice. Sediments from the bottom of the Norwegian Sea contain four layers of volcanic ash from Icelandic eruptions over the last 300,000 years. The layers indicate times when intense eruptions spewed out huge amounts of volcanic material, which settled on the ocean floor. All the layers fell within the short interglacial periods that punctuate the longer ice ages. Apparently, the ice sheets keep the volcanoes quiet by bearing down on the Earth's crust and putting weight on the magma chambers that feed the volcanoes. When the ice melts, the pressure on the chambers is lifted and the volcanoes erupt. In addition, it seems that the volcanoes are most active immediately after an ice age.

The explosive eruption of Tambora in Indonesia in 1815 exceeded any other known volcanic eruption during the entire Holocene epoch. It sent more volcanic ash into the upper atmosphere and obscured more sunlight than any volcanic eruption in the past 400 years. The eruption blew off the upper two-thirds of the mountain and cast some 25 cubic miles of debris into the atmosphere. It had a major impact on the climate throughout the Northern Hemisphere. By the summer of 1816, the ash had completely encircled the Earth. Temperatures dropped by as much as 7 degrees Fahrenheit in New England and 5 degrees or more in Europe. Ship captains in the North Atlantic reported sightings of large numbers of icebergs from a massive outbreak of Arctic ice (FIG. 5-11).

The event went down in history as the "year without a summer." Spring was late in New England. When it finally came and crops had begun to grow, a killing frost in June took all but the hardiest plants. When harvest was about to begin in the fall, a cold wave from the north brought widespread killing frosts and finished off

Fig. 5-11. Principal path of icebergs from their calving area in western Greenland.

those crops that had managed to survive the ordeals of the summer. The Europeans had it much worse because in addition to the cold summer, many parts of Europe had been ravaged by the Napoleonic wars that had ended the year before. The scarcity of food brought on insurrections and riots and eventually disease that killed over 100,000 people.

From 1645 to 1715, there was apparently a minimum of sunspot activity (FIG. 5-12) known as the Maunder Minimum for the English astronomer Walter Maunder who discovered it in 1894. It was blamed for a span of unusually cold weather in Europe and North America. The unusually low level of solar activity was also supported by a gap in Chinese naked-eye sunspot records for the same period.

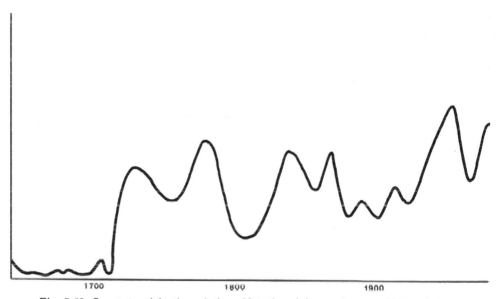

Fig. 5-12. Sunspot activity through time. Note the minimums between 1645 and 1715.

Ancient trees that were alive between 1645 and 1715 have been analyzed for ring growth (FIG. 5-13). They give an account of anomalous climatic activity that coincided with the Maunder Minimum of sunspot activity. The decrease in the Sun's luminosity, or brightness, might have been as much as 1 percent, causing an estimated cooling of 2 degrees throughout the world.

During the relatively warm and dry fifteenth and sixteenth centuries, major forest fires in North America appear to have occurred about every nine years. Over the next three centuries, however, during the cool period of the little ice age, forest fires were less frequent and less intense, occurring only about every 14 years. If greenhouse warming continues today, major forest fires such as those that devastated half of Yellowstone National Park during the summer and fall of 1988 (FIG. 5-14) might become more frequent. Heavy palls of smoke would blanket the sky and lower surface temperatures enough to bring on another little ice age.

(PHOTO BY L. E. JACKSON, JR., COURTESY OF USGS)

Fig. 5-13. Tree sample prepared for studies of annual ring growth.

Fig. 5-14. Forest fire that engulfed Yellowstone National Park in the summer and fall of 1988.

6

Terrestrial Causes of Glaciation

IT is necessary to study the many factors that influence the climate, in order to explain what caused the ice ages and the relatively short, warm interglacial periods. For instance, the large ice sheets caused additional cooling of the climate because the increased area of ice reflected more solar energy away from the Earth's surface. It would follow that if more heat was lost, the climate would become colder still and the glaciers would grow even larger. It would seem that with such a feedback mechanism in place glaciers would be difficult to stop. Yet the geologic record shows that dramatic changes in the climate did occur, which forced the retreat of the ice sheets. The difficult problem for scientists is finding a mechanism that regularly changed the climate over the past million years or so, turning the ice ages on and off, again and again.

VOLCANIC ERUPTIONS

In ice cores taken from Greenland, detectable traces of volcanic eruptions have been found that date back 10,000 years, near the end of the last ice age. Direct temperature measurements were made by comparing the changing proportions of oxygen-16 and oxygen-18 isotopes in the ice (see chapter 4). Because volcanic eruptions eject acid gases into the atmosphere periods of great volcanic activity are followed by periods of acid precipitation. The traces of acid in the ice conduct electricity more easily than ordinary ice and are detected by observing a drop in the electrical resistance through the ice core along its length. A comparison of the top layers of the ice with the historic record of volcanic eruptions shows good agreement. The acidity changes in the ice match up with other climate indicators such as tree rings, whose narrow bands of growth indicate colder or dryer seasons. They also correlate with written records of historic climate changes.

A volcanic eruption cloud contains particles that range in size from coarse ash to dust. The particles that remain airborne for long periods of time have the most affect on the climate, as does the nature of the dust and its location in the atmosphere. Volcanoes can eject ash and dust about 20 miles up into the troposphere and lower stratosphere. Finer particles can be blown to altitudes of 30 miles or more. It is the volcanic particles in the lower altitudes, however, that produce dense, long-lived dust clouds and seem to have the greatest influence on the climate.

Atmospheric scientists agree that it might not be the dust alone that blocks out heat from the Sun. Volcanoes also produce vast quantities of water vapor and gases, including sulfur dioxide, which reacts with water to produce sulfuric acid. These aerosols might penetrate the stratosphere like a fine mist and obscure sunlight. Aerosols are also transparent to outgoing infrared radiation. This heat loss would further cool the Earth. By expelling a combination of dust and aerosols into the atmosphere, large volcanic eruptions appear to have a significant impact on the climate.

Only part of the incoming solar radiation, however, is lost when it is reflected back into space by the volcanic dust cloud. Some of the Sun's energy warms the dust itself and some is scattered sideways. The energy still reaches the ground, but at an indirect angle. This sideways scattering of sunlight in the atmosphere is responsible for making the sky blue and for providing us with spectacular sunrises and sunsets.

If the total output of the Sun fluctuated by as much as 10 percent, as it might have in its early history, it would be disastrous for life on Earth because surface temperatures would respond with much larger fluctuations. This change could be modified, however, if it occurred at a time of increased volcanic activity. For example, a decrease of 5 percent in direct solar radiation after a major volcanic eruption would not cause the surface to cool by 10 degrees Fahrenheit, but by less than 1 degree. This is because the decrease in direct radiation is matched by an increase in scattered, indirect radiation caused by the volcanic dust cloud.

The effect on the climate from volcanic eruptions in the higher latitudes, such as those of Mount St. Helens, Washington, in 1980 (FIG. 6-1) and Mount Augustine, Alaska, in 1986, are not nearly as pronounced as those in the lower latitudes. This is because volcanic dust blasted into the stratosphere in the temperate zones tends to spread less. The dust from volcanic eruptions in the tropics is carried poleward by a high-altitude flow of air that originates in the tropics. It then concentrates in the higher latitudes. Since sunlight in the higher latitudes strikes the surface at a steep angle, the radiation must take a longer path through the dust cloud to the ground. This results in colder surface temperatures such as those experienced after the 1982 eruption of El Chichon in Mexico (FIG. 6-2)

The climatic effect of volcanic dust is dependent on the type of particles being released into the atmosphere, including the size of the dust particles and where they are concentrated in the atmosphere. Large dust particles in the troposphere could actually trap heat rising from the ground, which would otherwise escape into space, and warm the Earth. It is the smaller particles and aerosols that tend to allow the heat from the ground to escape and at the same time block the Sun's heat from reaching the Earth.

If the dust overlies light colored surfaces such as snow-packed areas, it absorbs

Fig. 6-1. The eruption of Mount St. Helens, Washington, on May 18, 1980.

more incoming solar heat than the surface. This results in less albedo and a net warming of the Earth. If the dust overlies a dark colored surface instead, such as a forest, it absorbs less heat than the surface creating more albedo and a net cooling effect. Volcanic dust that blocked heat from the Sun might have caused an increase in the glacial ice cover that resulted in the little ice age between the fifteenth and nineteenth centuries.

Fig. 6-2. The El Chichon Volcano, Mexico, after its explosive eruption in 1982.

Often, the longer a volcano remains dormant the more violent its eruption will be and the more particles it will pump into the atmosphere. There are at present several dozen candidate volcanoes, including some in the Cascade Range, that are expected to erupt at any time. Some climatologists argue that many volcanic eruptions occurring around the world in a short time span might inject enough dust into the upper atmosphere to cause the onset of an ice age. Their argument is supported by geological evidence of thin layers of volcanic dust buried in sediments that correlate with times of increased ice cover. Volcanic debris has also been found in various layers of Greenland and antarctic ice cores, providing strong evidence that frequent and violent eruptions accompanied the ice age.

THE CARBON CYCLE

Atmospheric carbon dioxide played a major role in the coming and going of the last ice age and possibly all previous ones as well. Soviet scientists have recovered a 7,000 foot long ice core from beneath their Vostok Station in eastern Antarctica. The core contains a continuous record of the temperature and composition of the atmosphere dating back 160,000 years. This spans a period from the last interglacial to the present one. Air bubbles trapped in the ice provide information on the carbon dioxide content of the atmosphere at the time the ice was laid down, while deuterium (an isotope of hydrogen) provides data on the temperature. What is highly significant is the fact that the level of carbon dioxide and the temperature kept in step throughout the entire period (Fig. 6-3).

During the last ice age, the level of carbon dioxide in the atmosphere was about 0.02 percent, which is only about half of what it is today. In addition to carbon dioxide, atmospheric methane, which is the second most important greenhouse gas and contributes to about one-quarter of global warming, also fluctuated during the last ice age. About 22,000 years ago, at the height of the glaciation, the level of methane was roughly half the pre-industrial level. Since that time, methane has risen nearly 250 percent due to man's activities. The reduction of methane during the ice age is blamed on the lower biological activity of wetlands and other habitats due to a colder climate.

The ocean played a major role in bringing down the level of atmospheric carbon dioxide. In the upper layers of the ocean, the concentration of gases is in equilibrium with the atmosphere at all times. These gases dissolve into oceanic waters mainly through the agitation of the surface waves. If the ocean were lifeless, much of its reservoir of dissolved carbon dioxide would enter the atmosphere, more than tripling the present amount. Luckily, the ocean is teeming with life. Marine organisms take up carbon dioxide, in the form of dissolved carbonates, to buiild their skeletons and other supporting structures. When the organisms die, their skeletons sink to the bottom of the ocean where they dissolve in the deep waters of the abyssal, which holds by far the largest reservoir of carbon dioxide.

In shallow water, the carbonate skeletons form deposits of carbonaceous sediments such as limestone and dolostone and are buried in the crust. The burial of carbonate in this manner is responsible for about 80 percent of the carbon deposited on the ocean floor. The rest of the carbonate comes from the burial of dead organic matter that has washed off the continents. Half of the carbonate is transformed back

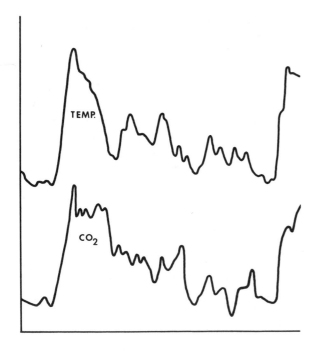

Fig. 6-3. The global temperature (top curve) and atmospheric carbon dioxide (bottom curve) have kept pace from 160,000 years ago to the present.

into carbon dioxide, which eventually escapes into the atmosphere. If it were not for this process, all the carbon dioxide would be taken out of the atmosphere in a mere 10,000 years. Photosynthesis would stop, and all life would end. Even a reduction of one-half the present amount of carbon dioxide would be enough to initiate a new ice age.

In this respect, life in the ocean acts like a pump that removes carbon dioxide from the surface and the atmosphere and stores it in the sea. The faster this biological pump works, the more carbon dioxide is removed from the atmosphere. This rate is determined by the amount of nutrients in the ocean, which respond to changes in ice volume. When the ice sheets began to melt about 16,000 years ago, quantities of atmospheric carbon dioxide rapidly increased. Then about 10,000 years ago, sea level rose from an influx of meltwater, which flooded continental shelves and removed organic carbon and nutrients. With reduced nutrients, the biological pump slowed down, allowing deep-sea carbon dioxide to return to the atmosphere.

Increased volcanic activity toward the end of the ice age played another important role in restoring the carbon dioxide content of the atmosphere. One of the most important volatiles in magma is carbon dioxide, which helps to make it flow easily. The carbon dioxide escapes from sediments when they melt after being forced into the mantle at subduction zones at the edges of crustal plates (FIG. 6-4). The molten magma along with its content of carbon dioxide rises to the surface to feed volcanoes that lie on the edge of subduction zones and at midocean ridges. When the volcanoes erupt, carbon dioxide is released from the magma and returned to the atmosphere, completing the cycle (FIG. 6-5).

MAGNETIC REVERSALS

Magnetic field reversals also have been cited as reasons for the ice ages. A reversal that occurred about 2 million years ago might have initiated the Pleistocene glacial epoch. Reversals in the magnetic field and excursions of the magnetic poles appear to correlate with periods of rapid cooling and the extinction of species. The Gothenburg geomagnetic excursion occurred about 13,500 years ago in the midst of a longer period of rapid global warming. It caused temperatures to plummet and glaciers to advance for 1,000 years, apparently in response to a weakened magnetic field.

The Earth's magnetic poles do not coincide with the axis of rotation. Presently they are offset at an angle of about 11 degrees. The poles are not stationary either, but slowly wander around the polar regions (FIG. 6-6). Since the turn of the century, the North Pole has crept about 30 feet toward eastern Canada. At this rate of polar wandering, Philadelphia could end up 10 degrees closer to the North Pole in 10 million years.

Major ice sheets that have waxed and waned over the last 2 million years might have been massive enough to cause the Earth to reorient itself and to produce the present polar wandering. This is because the Earth has a tendency to place most of its mass near the equator and large ice sheets are massive enough to upset the balance. There might also be a link between episodes of polar wandering and the tendency of the magnetic field to reverse itself.

Over the past four centuries, navigational charts have revealed two major trends

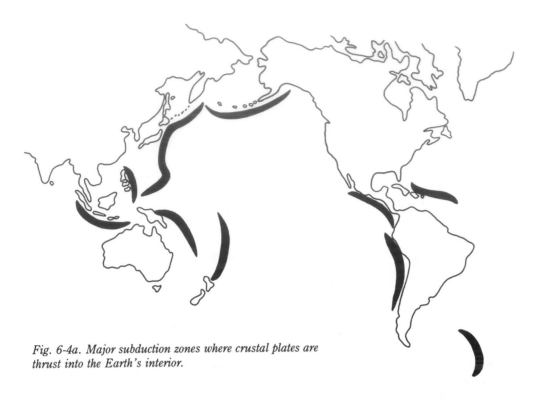

Fig. 6-4a. Major subduction zones where crustal plates are thrust into the Earth's interior.

(COURTESY OF NASA)

Fig. 6-4b. Topographic relief map of the ocean's floor generated from data from the Seasat satellite and depicting the subduction zones.

in variations in the magnetic field. The first is a slow, steady decrease in the intensity of the field. If it continues at such a rate, the magnetic field could collapse altogether in the not too distant future. The second variation is a slow westerly drift in irregular eddies in the field, amounting to one degree of longitude every five years. The drift suggests that the fluid in the outer metallic core, which generates the geomagnetic field, is moving at a rate of about 100 yards a day.

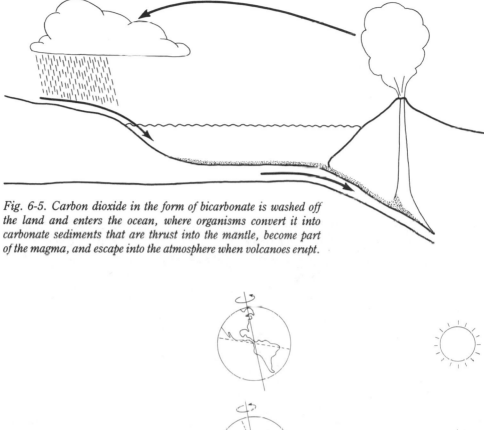

Fig. 6-5. Carbon dioxide in the form of bicarbonate is washed off the land and enters the ocean, where organisms convert it into carbonate sediments that are thrust into the mantle, become part of the magma, and escape into the atmosphere when volcanoes erupt.

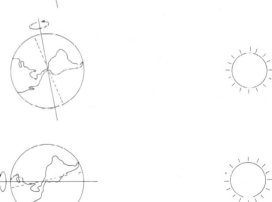

Fig. 6-6. Varieties of polar wandering and their effects on the climate.

The geomagnetic field protects the planet from dangerous cosmic radiation originating from the Sun and outer space. The solar wind, comprised of subatomic particles, stretches the magnetic field lines and creates the magnetosphere (FIG. 6-7), which protects the Earth from deadly ionizing radiation. The solar wind and the magnetosphere form a vast electrical generator, in which the interaction of magnetic fields and solar wind particles converts the energy of the solar wind's motion into electric-

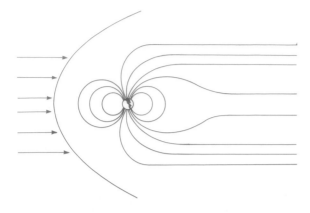

Fig. 6-7. The magneto-sphere shields the Earth from intense particle radiation in the solar wind.

Fig. 6-8. An aurora in the Southern Hemisphere as seen from Space Lab aboard the space shuttle.

ity. Where the magnetic lines of force descend toward the magnetic poles, charged particles from the solar wind smash into the ionosphere and light the northern or southern sky with an aurora (FIG. 6-8).

The magnetic field reverses in a highly irregular pattern that appears to be a random process. The magnetic field reverses on the average two or three times every million years. Over the last 170 million years it has reversed itself 300 times. The last time the magnetic field reversed itself was about 730,000 years ago. Since then, the Earth's climate has swung episodically between relatively warm times and generally cold periods, when thick ice sheets spread over the higher latitudes of the Northern Hemisphere.

Reversals also occurred during the Precambrian era and have been observed in all subsequent geologic time periods. There is no evidence that any one polarity has been favored over the other for long durations. The only exception possibly occurred during the Cretaceous period between 135 and 65 million years ago, when there appeared to be no reversals for more than 20 million years. Then there was an abrupt reversal at the end of the Cretaceous period, when the dinosaurs and many other species became extinct.

Reversals in the magnetic field might occur along with reversals of convective currents in the core, which are responsible for generating the magnetic field through the so-called dynamo effect. (A dynamo is a device that generates an electrical current by rotating a conducting medium through a magnetic field. The faster the rotation the greater the current and consequently the greater the magnetic field.) Reversals in the convective currents might result from fluctuations in the level of turbulence in the core caused by heat loss at the mantle-core boundary, and from the progressive growth of the inner core, which provides the gravitational energy to power the dynamo. Changes in core pressure brought on by tectonic events, ice ages, and large asteroid impacts also could cause fluctuations in the core's turbulence.

After a period of up to a million or more years of stability, the strength of the magnetic field gradually decays over a short time span, probably no more than 10,000 years. It then abruptly collapses, reverses itself, and slowly builds back to its normal strength. This process might take upwards of 1,000 years before the field regains its full magnetic intensity.

It has long been suspected that there is a correlation between changes in the magnetic field and events that have taken place on the surface of the Earth. In many respects, a comparison of the reversals with known ups and downs in the climate shows a striking agreement. Magnetic reversals that occurred 2.0, 1.9, and 0.7 million years ago coincide with unusual cold spells, as indicated by the ratio of carbon and nitrogen in ancient lake-bed sediments. High ratios indicate dwindling nitrogen levels due to shrinking populations of algae and plankton brought on by colder climates.

When the Earth lets down its magnetic shield during a geomagnetic reversal, the climate cools. The effect is similar to that of intense bursts of solar activity (FIG. 6-9), which also tend to cool the Earth. Variations in magnetic field intensity over the the past several hundred thousand years closely agree with variations in surface temperatures. When the field is weak, more cosmic rays are allowed to penetrate and warm the lower atmosphere. This causes a thermal disturbance in the atmosphere, which in turn can affect the climate.

When the magnetic field strength is down, the atmosphere is exposed to solar wind and high intensity cosmic radiation. The increased bombardment might influence the composition of the upper atmosphere by making more nitrogen oxides. This

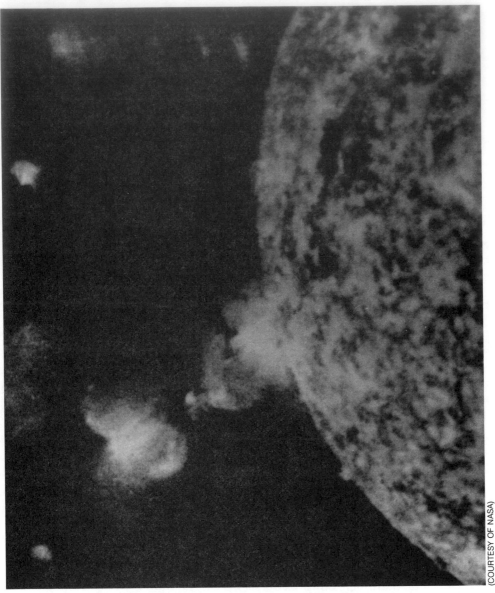

Fig. 6-9. Solar eruption billowing forth clouds of particles.

in turn, might block out sunlight and possibly alter the climate enough to bring on an ice age.

Periods of intense solar activity appear to have an affect on the weather. Any reduction in the magnetic shield would have a similar effect. Those latitudes that are particularly sensitive to changes in the Sun's activity would be similarly affected by a decrease in the magnetic field. Since magnetic reversals coincide with extinctions, there is also a possible link between the magnetic shield, atmospheric composition, and even the evolution of life itself.

84   TERRESTRIAL CAUSES OF GLACIATION

PLATE TECTONICS

It is becoming more apparent that plate tectonics and continental drift have played a prominent role in climate and life since the early stages of the Earth, possibly as far back as 2.7 billion (TABLE 6-1). Changes in continental configurations brought on by movable crustal plates greatly affected global temperatures, ocean currents, biological productivity, and many other factors of fundamental importance to the Earth.

The positioning of the continents helped determine climatic conditions. When most of the land huddled near the equatorial regions, the climate was warm. When lands wandered into the polar regions, however, the world became covered with ice. During times of highly active continental movements, there was greater volcanic activity, especially at subduction zones and mid-ocean ridges. The amount of volcanism could have affected the composition of the atmosphere and the rate of mountain building, which ultimately affected the climate.

All the lands were welded into a single large continent, called pangaea, near the tropics about 230 million years ago. This allowed more of the Sun's heat to be absorbed, which in turn contributed to higher global temperatures. Oceans that existed in the higher latitudes were less reflective than land and absorbed more heat, which further moderated the climate. Because there was no land located in the polar regions to interfere with the movement of warm ocean currents both poles remained ice free year round. There was no large variation in temperature between the high latitudes and the tropics.

During the breakup of pangaea 180 million years ago (FIG. 6-10), the climate, particularly during the Cretaceous period, was extremely warm. Average global temperatures were 10 to 25 degrees Fahrenheit warmer than they are today. When the continents drifted toward the poles at the end of the Cretaceous period, however, they disrupted the transport of poleward oceanic heat and substituted reflective, easily chilled land for absorptive, heat-retaining water. As the cooling progressed, the land accumulated a greater reflective surface of snow and ice, which further lowered global temperatures.

Land existing near the poles is often the cause of extended periods of glaciation because high-latitude land has a higher albedo and lower heat capacity than the surrounding seas. This encourages the accumulation of snow and ice. The more land area in the higher latitudes, the colder and more persistent is the ice. This is especially true when much of the land is at higher elevations. Taking land away from the tropics and replacing it with ocean waters also has a net cooling effect because land absorbs more of the Sun's heat, while the oceans reflect it back into space. When the land area above a certain latitude increases and is covered by steady snow falls, a permanent polar glacial climate may result.

Once the glaciers are in place, the high reflectivity of snow and ice tends to perpetuate them and sustain a glaciation, even if the once high land was to sink to the level of the sea because of the weight of the overlying layers of ice. Some scientists believe that the increasing weight of ice can actually squeeze magma out of the Earth and cause an increase in volcanic activity. The greater the volcanic activity, the greater the amount of ash cast up into the atmosphere. This further cools the Earth, which in turn increases glaciation.

The congregation of land in one area also affects the shapes of the ocean basins.

	TABLE 6-1. The Drifting of the Continents	
AGE × MILLION YEARS	GONDWANALAND	LAURASIA
Quaternary 3		Opening of Gulf of California
Pliocene 11	Begin spreading near Galapagos Islands	Change spreading directions in eastern Pacific
	Opening of the Gulf of Aden	
		Birth of Iceland
Miocene 25		
	Opening of Red Sea	
Oligocene 40		
	Collision of India with Eurasia	Begin spreading in Arctic Basin
Eocene 60		Separation of Greenland from Norway
	Separation of Australia from Antarctica	
Paleocene 65		Opening of the Labrador Sea
	Separation of New Zealand from Antarctica	Opening of the bay of Biscay
	Separation of Africa from Madagascar and South America	Major rifting of North America from Eurasia
Cretaceous 135		
	Separation of Africa from India, Australia New Zealand and Antarctica	
Jurassic 180		Begin separation of North America from Africa
Triassic 230		
Permian 280		

Calving of blue ice at the terminus of Worthington Glacier near Valdez, Alaska.

Terminal moraine of a cirque glacier in the San Juan Mountains, Colorado.

A glacier-hung mountain wall northeast of Mt. Sefton, South Island, New Zealand.

A glacial valley near Browns Pass in Glacier National Park, Montana.

Arapaho Glacier, Boulder County, Colorado.

Knob and kettle topography near Red Glacier, Alaska.

A crevasse in the Wilkins Mountains on the Antarctic Peninsula.

An iceberg near Hallett Peninsula, Antarctica.

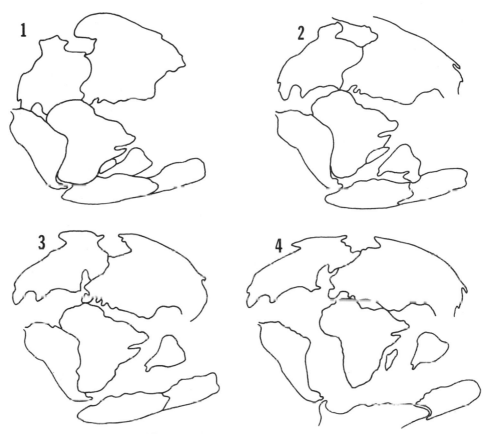

Fig. 6-10. The breakup of Pangaea 225 (1), 180 (2), 135 (3), and 65 (4) million years ago.

The ocean bottom influences the amount of heat that is carried by ocean currents from the tropics to the poles (TABLE 6-2). When Antarctica separated from South America and Australia some 40 million years ago and moved to the South Pole, a circumpolar antarctic ocean current was established (FIG. 6-11). This current isolated the frozen continent and prevented it from being warmed by poleward flowing waters that originated in the tropics.

The Arctic Ocean, on the other hand, is the only ocean in the world that is practically landlocked. The land stops warm, tropical ocean currents from reaching the North Pole and melting the polar ice cap. If the Bering Strait were to become totally blocked with ice, this would effectively keep the warm Pacific current from entering the Arctic Ocean, which could keep it frozen and unnavigable year round. This condition could initiate the spread of glaciers across the northern lands.

Changes in ocean water salinity also affect the formation of sea ice (FIG. 6-12). If less fresh water entered the Arctic Ocean, its salinity would increase. The saltier, heavier water would sink to the bottom and cause the ocean to oveturn. This would force colder, less salty water to the surface where it would freeze into pack ice. The amount of sea ice in the Arctic varies from over 3 million square miles in the summer to nearly twice that amount in the winter. If none of the ice were allowed to melt in

TABLE 6-2. History of the Deep Circulation in the Ocean

AGE	EVENT
> 50 million years ago	The ocean could flow freely around the world at the equator. Rather uniform climate and warm ocean even near the poles. Deep water in the ocean is much warmer than it is today. Only alpine glaciers on Antarctica.
35–40 million years ago	The equatorial seaway begins to close. There is a sharp cooling of the surface and of the deep water in the south. The Antarctic glaciers reach the sea with glacial debris in the sea. The seaway between Australia and Antarctica opens. Cooler bottom water flows north and flushes the ocean. The snow limit drops sharply.
25–35 million years ago	A stable situation exists with possible partial circulation around Antarctica. The equatorial circulation is interrupted between the Mediterranean Sea and the Far East.
25 million years ago	The Drake Passage between South America and Antarctica begins to open.
15 million years ago	The Drake Passage is open; the circum-Antarctic current is formed. Major sea ice forms around Antarctica which is glaciated, making it the first major glaciation of the Modern Ice Age. The Antarctic bottom water forms. The snow limit rises.
3–5 million years ago	Arctic glaciation begins.
2 million years ago	An Ice Age overwhelms the Northern Hemisphere.

the short warm season, it could continue its expansion onto the continents, bringing with it the full force on an ice age.

The polar ice caps act as major heat sinks and help drive the large-scale atmospheric and oceanic circulation systems (FIG. 6-13), which have a tremendous effect on the global climate. The moisture-laden hot air that rises from the tropics is forced to move poleward by cooler, heavier air masses, which are indirectly cooled by the polar ice. The polar regions therefore play a significant role in major long-range alterations in global climate patterns and have a considerable impact on the mid-latitude climate.

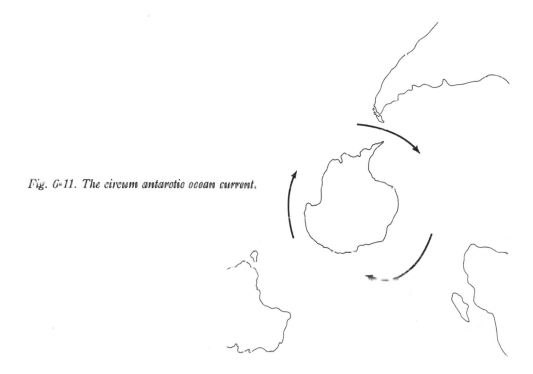

Fig. 6-11. The circum antarctic ocean current.

As the ice cover fluctuates from year to year, so does its effect on the weather. If ice cover is reduced, the amount of open ocean and subsequent evaporation is increased, which produces more cloud cover. More clouds lower the temperature and allow the ice cover to increase. This reduces evaporation and cloud cover and completes the cycle. Thus, the ice caps provide an effective feedback mechanism that moderates the climatic temperature.

If the ice caps were to suddenly disappear, the Earth's climate would change dramatically because of the breakdown in the circulation system that transfers heat away from the tropics to other parts of the world. This might also make the tropics too hot to be habitable. The poles would not stay ice free for very long. Because they would no longer receive warmth from the tropics glaciers would quickly grow beyond their present size and overrun the continents.

EXPANSION OF THE FORESTS

The glacial episode that occurred toward the end of the Paleozoic era, some 260 million years ago, might have been triggered by the spread of forests. The massive forests withdrew large amounts of carbon dioxide from the atmosphere and buried it in the sediments as coal. This weakened the greenhouse effect and caused the climate to cool. The burial of carbon dioxide in the crust might have been the key to the onset of all major ice ages since life evolved on Earth.

A radical change in vegetation occurred during the middle of the Cretaceous period with the introduction of angiosperms, or flowering plants. They were distributed worldwide by the close of the Cretaceous period and today include about a quarter

Fig. 6-12. Grounded growlers in the Arctic Ocean.

of a million species of trees, shrubs, grasses, and herbs. The increase in angiosperms even might have contributed to the extinction of the dinosaurs and certain marine species at the end of the Cretaceous period because the plants absorbed large quantities of carbon dioxide, which lowered global temperatures.

At the end of the Cretaceous period, polar forests existed at latitudes up to 85 degrees, well beyond the present-day tree line (FIG. 6-14). The most remarkable ex-

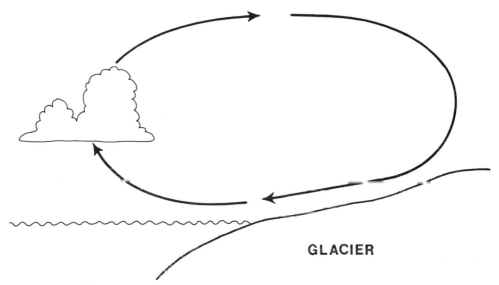

GLACIER

Fig. 6-13a. The ice caps drive atmospheric circulation.

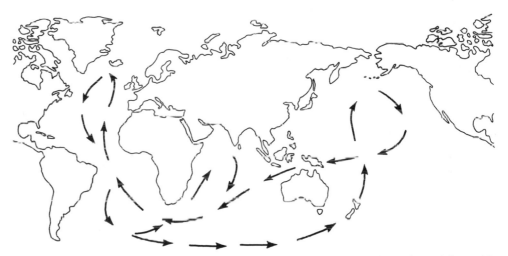

Fig. 6-13b. The ocean heat transport system distributes warm water to colder regions of the world.

ample of this is a well-preserved fossil forest on Alexander Island, Antarctica. The strange location of this forest might have been the result of a decrease in the angle of the Earth's axis to as little as 5 degrees. A small change in the degree of tilt would have diminished sharply the difference in the seasons and allowed more evenly spaced sunlight throughout the year.

Even though such a decrease in the Earth's tilt would spread sunlight more evenly, it would also reduce the amount of mean annual heat. It seems more likely that the tilt was not much different than it is today. Instead, the trees probably adapted mechanisms for intercepting the maximum amount of sunlight during a period when global temperatures were considerably warmer than they are today.

The cone-bearing plants, which were prominent during the Mesozoic era, occupied a secondary role during the Cenozoic era. Tropical vegetation, which was widespread during the Mesozoic, withdrew to narrow regions around the equator because of a colder, drier climate, that resulted from a general uplifting of the landmasses and the draining of interior seas. Forests of giant hardwood trees that once grew as far north as Yellowstone National Park are now replaced by conifers, which indicates that today's climate is cooler.

Fig. 6-14. The arctic tree line, north of which no trees grow.

7

Celestial Causes
of Glaciation

THE shape of the Earth's orbit and the precession and degree of tilt of its spin axis influences climate. Even the alignment of the planets, which cause gravitational effects on the Sun and Earth, affects the Earth's climate. The Solar System passing through a vast dust cloud as it travels around the Galaxy might also influence the amount of sunlight the Earth receives. These cyclical cosmic influences might be responsible for the extermination of species and the great ice ages.

GALACTIC DUST CLOUDS

The Solar System circles around the galaxy. It completes one revolution approximately every 200 million years. Two or three times a century, somewhere in one of the spiral arms of the Galaxy, a giant star explodes and becomes a supernova. The expanding supernova injects huge amounts of dust and debris into the Galaxy. The Solar System might enter such a dust cloud every 100 million years or so. If the Solar System happened to pass through relatively dense regions of this intergalactic dust cloud, the material falling into the Sun might affect its energy output. The dust also might block out the Sun and alter the amount of sunlight the Earth receives.

Passage through a dust cloud could take several million years. Such a dust cloud might have been responsible for some of the earlier glacial epochs, which lasted for a similar length of time. However, this would be too lengthy a time period to explain the relatively short ice ages of the past 2 million years. It also seems doubtful that passing through a dusty arm of the Galaxy would have caused the continuous temperature decrease from the Cretaceous period to the present.

There does appear to be a cloud of interstellar gas streaming through the Solar System (FIG. 7-1), however. In its lifetime, the Sun has passed through at least 100 dense gas clouds. During the 100,000-year traverse through the interstellar cloud, the Earth would acquire a large amount of molecular hydrogen. The hydrogen would react with chemical constituents in the upper atmosphere to produce water vapor, which condenses into clouds. The clouds would reflect a large amount of solar radiation, which could lower surface temperatures several degrees. If this climate were sustained for several thousand years, it could trigger an ice age.

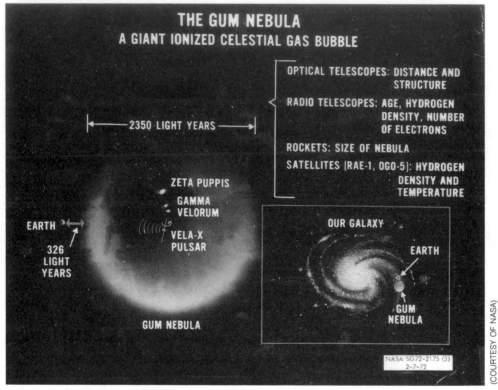

Fig. 7-1. *The Gum Nebula, which is 1,500 light years from Earth, was probably produced by a supernova.*

The Sun's journey through dense clouds located at the midplane of the Galaxy could also reduce the Earth's insolation and initiate climatic changes. However, there is no evidence to suggest that the dust cloud is sufficiently dense to block out the Sun during each passage through the midplane, which could take upwards of several million years to complete. As the Sun reaches the upper or lower regions of the Galaxy, the Earth could be exposed to higher levels of cosmic radiation from an exploding supernova. This would ionize the upper atmosphere and produce a haze that would block out sunlight and cool the planet.

SOLAR INPUT

The total amount of solar energy intercepted by the Earth, spread uniformly across the surface and averaged over a year's time, is known as the average solar input. For centuries, astronomers have referred to a solar constant. This means that the total amount of solar energy impinging on the Earth has remained fairly steady through time. The solar constant depends on the Sun's luminosity, or brightness, and the Earth's orbit. Luminosity depends on the Sun's size and surface temperature. Reducing the solar constant by only a few percent is sufficient to bring on a major ice age.

Although the Sun's average energy output seems fairly steady over the short run, it has actually been steadily increasing. About 4 billion years ago, the Sun was approximately 8.5 percent smaller and its luminosity was 3 to 4 percent less than it is now. This implies that the solar constant was as much as 30 percent less than it is at present, or equivalent to the solar energy Mars receives today. The Sun is not expected to get much hotter during the next 4 billion years. It will continue to grow in size, however, as it depletes the hydrogen fuel in its core. About 5 billion years from now, the Sun will begin to expand and eventually encompass the orbit of Mercury. Long before that would happen though, all life on Earth would be extinguished by the blazing heat from the Sun.

Solar luminosity is controlled by thermonuclear reactions taking place in the Sun's core and the properties of the gasses in the outer layers (FIG. 7-2). Energy created by the conversion of hydrogen into helium requires a constant depletion of hydrogen and produces an increase in luminosity over a very long time period.

The neutrino problem calls attention to the fact that the nuclear reactions taking place in the Sun are not as straightforward as was once thought. Since 1968, scientists from the Brookhaven National Laboratory have been operating a solar neutrino collector, which is buried deep in a South Dakota gold mine. They have only been able

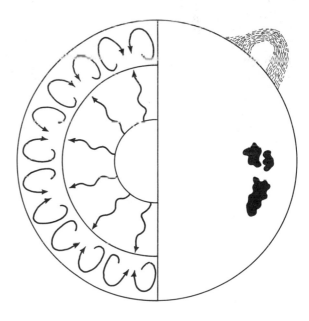

Fig. 7-2. The nuclear reactions in the interior of the Sun control actions on the surface, including sunspots and solar flares.

to detect about one-third the number of neutrinos that they expected to find. One possible reason for this observation is that the Sun's core temperature, thought to be about 15 million degrees Celsius, is less than the theoretical amount and thus produces less neutrinos. A lower core temperature, however, certainly would give the Sun a different luminosity than what is observed now. In addition, if the core were actually cooling down, the Sun's output would diminish considerably.

A long-term change in luminosity cannot be detected from Earth or by orbiting satellites. However, small, short-term fluctuations are detectable. Many meteorologists and climatologists believe these changes to be sufficient to produce variations in the climate. These small fluctuations in solar output come in regular intervals of 22 years known as the solar cycle. Coinciding with the solar cycle is an 11-year sunspot cycle. There also might be cycles of 90 and 180 years or more, which might correlate with the ice ages.

It is now well established that the sunspot cycle is not strictly periodic, but ranges from about 9 to 14 years in duration. Other solar activity including solar flares, solar cosmic rays, ultraviolet radiation, and x-rays varies directly with the sunspot cycle. Moreover, the polarity of the Sun's magnetic field reverses from cycle to cycle producing a double cycle of roughly 22 years.

In the course of each sunspot cycle, sunspots tend initially to appear at the high solar latitudes and then move progressively closer to the solar equator. This observation led to the discovery that the Sun does not rotate as a solid body. Instead, the sunspots at the equator rotate faster than those at higher latitudes. Sunspots are also associated with magnetic fields that are several thousand times stronger than the magnetic field on the Earth's surface. The Sun's magnetic field is thought to be generated by the difference in rotation speeds between the rapidly spinning core and the upper gas layers.

During a sunspot maximum, when large numbers of sunspots mar the Sun's surface (FIG. 7-3), the activity of the Sun increases. It did this by as much as 0.4 percent between 1976 and 1979. The Sun's activity has been steadily increasing since 1985, and may be responsible for the severe droughts of the late 1980s. When there are no sunspots, the Sun can cool by as much as 1 percent. Such a condition might have existed during the little ice age, a span of unusually cold weather in Europe and North America from the sixteenth to the early nineteenth century.

The Maunder Minimum, a 70-year cessation of sunspot activity from 1645 to 1715, was linked to the coldest part of the little ice age. The discovery of this period of an unusually low level of solar activity is supported by a gap in East Asian naked-eye sunspot records for the period between 1645 and 1715. Tree rings that formed between 1645 and 1715 show a period of very unusual solar behavior that matches closely that of the Maunder Minimum. The wider the tree rings, the better the climate was during the time when each ring was formed. Evidence of another Maunder-type minimum that occurred during the Ming dynasty between 1400 and 1600 has also been discovered. This might have sparked the beginning of the little ice age.

Lake bed sediments, in southern Australia, called varves, show distinct bands of mud and silt. Dating to the Precambrian ice age about 680 million years ago, these bands are of varying widths and have numerous lighter bands that are interspersed with darker bands. When the lake bed sediments were being laid down, much of Australia was covered by thick ice sheets (FIG. 7-4).

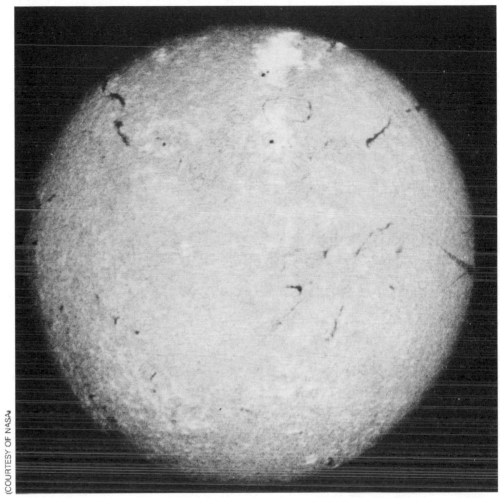

Fig. 7-3. The surface of the Sun showing sunspots and solar flares.

Fig. 7-4. The Late Precambrian ice sheet overlying
Australia.

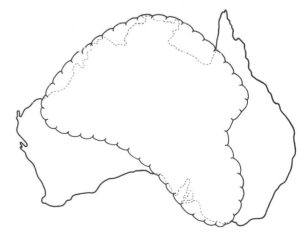

Increased solar activity eventually caused a corresponding increase in the temperature on Earth. This entailed a greater annual discharge of glacial meltwaters and the deposition of thicker, darker layers of sediment. The varves appear to mimic the solar cycle, which has periods of approximately 11, 22, and 90 years. Periods of 145 and 290 years also appear to match modern rhythms in some tree ring climate records. This correspondence between sunspot activity and varve thickness provides a strong argument for a link between solar activity and terrestrial climate 680 million years ago.

Newly discoverd sediments, near the other deposits north of Adelaide, Australia, with far thicker sedimentary cycles that contain 14 or 15 laminations per cycle might better represent the lunar tidal cycle. Today the lunar cycle has a period of 18.6 years, indicating that the Moon was much closer to the Earth during the late Precambrian period.

ORBITAL MOTIONS

All orbital motions of the Earth are identified by three elements: the geometry of the orbit, the precession of the equinoxes, and the tilt of the spin axis. None of the orbital elements will affect the total amount of solar radiation reaching the Earth during a year's time. They only alter the amount of solar energy reaching certain latitudes in certain seasons. Thus, one area might have cold winters and hot summers during one cycle and mild winters and cool summers during another.

The Earth's orbit around the Sun varies from an almost perfect circle to an ellipse. It takes about 100,000 years to complete one cycle. When orbiting the Sun in a circle, the Earth maintains a constant distance of 93 million miles during all seasons. The amount of insolation is the same throughout the year. During the ellipse phase, the Earth is closer to the Sun during one season, making it warmer, and further away during the opposite season, making it cooler. Presently, the Earth is in an elliptical orbit and its perihelion, or closest approach to the Sun, occurs in early January. Its aphelion, or farthest recession from the Sun, occurs in early July. The difference between perihelion and aphelion is 3 million miles. The total insolation is 7 percent less during the northern summer than it is during the winter.

This orbital cycle might explain the waxing and waning of the great ice ages every 100,000 years for the last million years or so (FIG. 7-5). It is still a mystery,

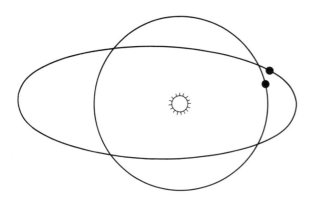

Fig. 7-5. The Earth's orbit stretches from nearly circular to elliptical and back again in about 100,000 years, making the Earth's distance from the Sun vary by about 11 million miles.

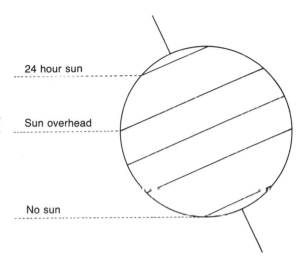

Fig. 7-6. The effect of the tilt of the Earth's axis on the seasons. Shown is summer in the Northern Hemisphere.

24 hour sun

Sun overhead

No sun

however, why the weakest cycle, which is responsible for less than one-half a percent of variation of insolation, produces the largest climatic changes. Perhaps it only initiates changes in climate that then amplify and reinforce each other. The Earth might also produce the 100,000-year ice age cycle on its own without orbital influences.

The Earth's shorter tilt and wobble cycles change the amount of sunlight the Northern Hemisphere receives in summer by as much as 20 percent. The Earth's spin axis is tilted at an angle of 23.5 degrees in relation to the plane of its orbit. The Sun and Moon exert a gravitational pull on the spinning Earth, making its axis wobble, or precess, like a toy top. The axis describes a cone in the heavens that precesses clockwise, which is the opposite direction of the Earth's rotation. It takes 21,000 years to complete one precession cycle. This means that about 10,000 years ago, at the end of the last ice age, Vega was the North Star and the seasons were reversed. In another 10,000 years, the Earth will again be tilted in the opposite direction. Those constellations presently seen only in the Southern Hemisphere will then be seen only in the Northern Hemisphere.

The tilt of the Earth's axis has varied from 22 to 24.5 degrees. Changing the angle of tilt shifts the position of the Sun during the seasons (FIG. 7-6). The greater the tilt, the greater the difference between summer and winter temperatures. If the Earth's axis were not tilted and were perpendicular to the plane of the ecliptic, the planet would have no seasons. On the other hand, if the axis were steeply tilted, the Earth would experience large temperature variations from one season to the next. Even slight changes in the degree of tilt can cause large climatic effects. The Earth completes one full tilt cycle every 41,000 years. Since the end of the last ice age, the degree of tilt has been steadily decreasing, which could produce cooler summers and warmer winters.

Variations in orbital motions combine to produce overall changes in the pattern of solar radiation falling on the Earth. If the amount of summer sunshine in the Northern Hemisphere should drop below a certain level and produce cool summers, the volume of the ice sheets would grow in proportion to the sunlight deficit. Snow that had fallen the winter before would fail to melt and the following winter's snow would pile on top of it. If this should continue without disruption for several years, the process of changing from an interglacial age to a full ice age could take place in as little as 200 years.

If a large portion of the Northern Hemisphere were to become snow bound all year, the snow's high albedo would reflect back into space solar radiation that otherwise would have warmed the Earth. The ice sheets would become self-perpetuating—one reason why ice ages linger for upwards of 100,000 years. Interglacial periods, which last only about 10,000 years, are short-lived events in the Earth's history because they are more susceptible to climatic changes. Once an interglacial is established, it can exist only as long as the darker, snow-free ground absorbs warmth from the Sun and prevents the glaciers from growing.

In 1941, the Yugoslav mathematician and astronomer Milutin Milankovitch, after calculating the amount of insolation at various latitudes over the seasons, proposed the theory of orbital variations (FIG. 7-7). His thesis was that cool summers instead of severe winters were all that would be necessary to initiate an ice age.

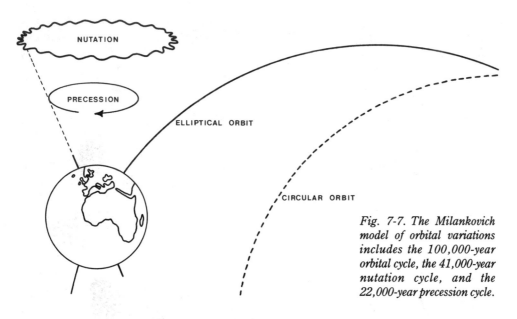

Fig. 7-7. The Milankovich model of orbital variations includes the 100,000-year orbital cycle, the 41,000-year nutation cycle, and the 22,000-year precession cycle.

In order to prove the Milankovitch model, some fairly accurate dates have to be applied to the major ice ages. One means of determining the age of glacial periods is by dating the fluctuations in sea level (FIG. 7-8). During an ice age, the sea level lowers appreciably because a great deal of water is locked up in the glaciers. Coral only grows near sea level (FIG. 7-9), so fluctuations in the level of the sea can leave a staircase of coral growth that corresponds with periods of glaciation. The age of the ancient coral is determined by radiometric dating techniques that calculate the degree of decay of certain radioactive elements (FIG. 7-10). The ages of the coral terraces tend to be some 20,000 years apart, comparing favorably with the cycle of variations in the tilt of the Earth's axis.

The Deep Sea Drilling Project, sponsored by the National Science Foundation,

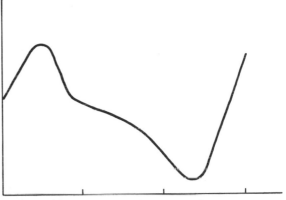

Fig. 7-8. Relative sea level during the last 150,000 years.

Fig. 7-9. Coral reef off of Bikini Atoll.

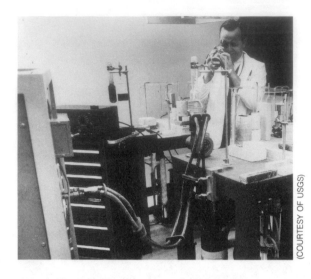

Fig. 7-10. Potassium-argon dating of sample.

has yielded evidence of a 400,000-, 100,000-, and 40,000-year cycle by analyzing the calcium carbonate content of ocean bottom sediments. Climate-related changes in the dissolving power of sea water, the level of the sea, circulation patterns, and the rate of erosion of the continents all affect the proportion of calcium carbonate in the sediments. When tiny marine organisms, which use the carbonate to make their shells, die and are buried in the sediments, they take to their graves an indicator of the climate during their lifetime. By comparing the content of the heavy isotope oxygen-18 (O-18) and oxygen-16 (O-16) in the fossilized shells of these creatures, scientists have a direct means of determining the Earth's climate.

When seawater evaporates in a warm climate, water molecules composed of both oxygen isotopes are evaporated and eventually land as snow at the poles. During cooler climates, however, water molecules that are weighed down with O-18 are less apt to evaporate and thus become concentrated in the seawater. Therefore, the more O-18 locked up in the shells of marine organisms, the colder the climate. Chemical analysis of the Greenland and Antarctic ice sheets have also yielded a direct means of measuring temperature by comparing the ratio of O-18 to O-16. One striking feature of this analysis is how clearly it follows the 41,000-year cycle, yielding strong support for the Milankovitch model of glaciation.

The Earth's orbital variations might have modulated the climate for hundreds of millions of years. Repetitive sequences that correspond to cycles of varying lake depth have been found in 200-million-year-old banded lake bed sediments from the Newark Basin in northern New Jersey. The cycles closely resemble the periods of precession of the Earth's axis and variations in the shape of its orbit.

When the sediments were dated, they yielded periods of 25,000, 44,000, 100,000, 125,000 and 400,000 years. These compared reasonably well with the present orbital periods of 21,000, 41,000, 95,000, 123,000, and 413,000 years. The cycles were also verified by studying sediment samples from throughout the world. From these examples, it appears that orbital variations operated on the climate for most of the Earth's history.

PLANETARY ALIGNMENT

It is known that the gravitational forces of the Sun and planets cause the Earth's spin axis to wobble. There is even speculation that the solar cycle is regulated by the gravitational forces of the planets, particularly the inner planets and Jupiter. Throughout this century, the alignment of the inner planets on one side of the Sun seemed to produce different sunspot numbers than when the planets were scattered around the Sun. Jupiter's orbit is also roughly the same as the 11-year sunspot cycle. Since Jupiter is the largest of the planets, it would seem to have the greatest gravitational affect on the Sun.

When the planets line up on the same side of the Sun, their gravitational pull raises tides on the Sun's surface just as the Sun and Moon raise tides on the oceans (FIG. 7-11). The tides are probably small because of the distance of the planets and the small gravitational forces involved. The years of sunspot maximum and minimum since 1800, however, coincide closely with the Sun's tidal maximum and minimum.

SUN

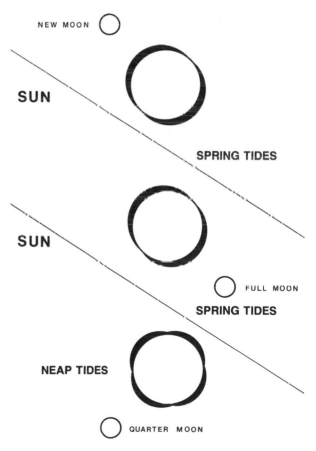

Fig. 7-11. The ocean tides are formed by the gravitational pull of the Sun and Moon.

This could have far-reaching implications for the prediction of sunspots and the climate.

The alignment of the planets also has been thought to have a direct influence on the Earth's weather and on sunspot activity. Ancient Chinese astronomers were the first to recognize this alignment, which they called a planetary synod. Recently, Chinese researchers have found that the planetary alignments have affected the weather for the past 3,000 years. The 180-year synods also coincide with the ups and downs in the temperature record of the Greenland ice cores and match the 180-year cycle of solar activity.

The planets rotate around the Sun at different rates. The inner planets complete their orbits several times faster than the outer planets. There are nine separate and independent orbital periods. It takes about 180 years for the Earth to be on one side of the Sun when the rest of the planets are bunched together on the other side. This displaces the gravity center of the Solar System and stretches the Earth's orbit by nearly a million miles. As it is further away from the Sun, the Earth cools by a small amount for several years. The last planetary synod occurred in October 1982. Chinese astronomers have predicted that this could initiate a cold spell that might last as long as half a century.

METEOR BOMBARDMENT

One of the most fascinating theories for sudden mass extinctions, which seem to have occurred roughly every 26 million years over the past 600 million years, deals with a hypothetical companion star of the Sun. Most stars in our galaxy have one or more companion stars that are bound by a common center of gravity, making them orbit each other. The Sun's supposed companion star was named Nemesis in honor of the Greek goddess who dished out punishment on the Earth.

Surrounding the Sun about a light-year away are trillions of comets in what is known as the Oort Cloud. Nemesis, which is too dim and too distant for astronomers to see, is believed to be circling the Sun in a highly elliptical orbit. Every 26 million years or so, the Sun's companion star swoops out of the heavens and passes close by the Oort Cloud. Its gravitational attraction scatters a large number of comets in all directions. Many of these shower down on the Earth. The impacts greatly disrupt the climate and might have disrupted the Earth's magnetic field in the past, causing it to reverse. Comets might also have been responsible for massive volcanic eruptions.

A search for meteorite craters is a difficult task because erosion has long since erased them. On the Moon, Mars, and Mercury, however, the impacts are quite evident, and numerous (FIG. 7-12). Because the craters tend to overlap each other any regular pattern that could be recognized is probably lost. Fortunately, there are still a few remnants of the ancient craters left on Earth. For example, the Manicouagan River and its tributaries in Quebec create a reservoir around a roughly circular structure some 60 miles across (FIG. 7-13). The structure is composed of Precambrian rocks that have been reworked by shock metamorphism, which could have been caused by an impact from a large celestial body. The impact crater was formed about 210 million years ago, which coincides with the end of the Triassic period, when mass extinctions occurred. This date is eight times the orbital period of Nemesis.

The origin of the meteorites has been a long-standing puzzle. The most accepted

Fig. 7-12. A large lunar crater viewed from Apollo 17.

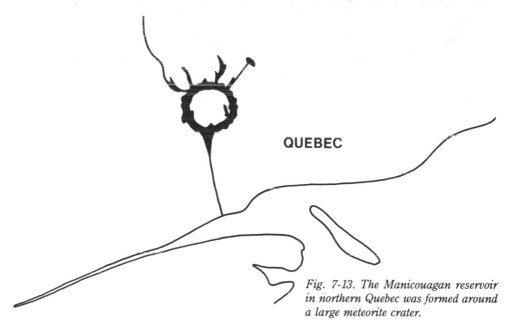

QUEBEC

Fig. 7-13. The Manicouagan reservoir in northern Quebec was formed around a large meteorite crater.

theory is that they came from a jumble of asteroids that existed in a wide belt between Mars and Jupiter. Some asteroids might be the rocky cores of comets that settled into orbit around the Sun and lost their coat of ice. Certain rare meteorites might be pieces of the Martian crust blasted out by large asteroid impacts (FIG. 7-14). Even pieces of the Moon are thought to make their way to Earth from major collisions.

The asteroids range in size from a chunk of rock hundreds of miles wide to small grains called micrometeorites. No one has yet been able to explain how these rock fragments managed to get into orbits that crossed our planet's path. Apparently, the asteroids behave themselves fairly well and run in nearly circular orbits for as long as a million years or more. Then, for unknown reasons, possibly due to a passing comet or the gravitational attraction of Jupiter, their orbits suddenly stretch to become so elliptical that a few of them actually fall on the Earth.

Meteorite showers are a more common occurrence than most people realize. Every day thousands of meteors rain down on the Earth. Occasionally, meteor showers can involve hundreds of thousands of stones. Upwards of 1 million tons of meteorite material is produced annually. Luckily, most meteors burn up upon entering the atmosphere and their ashes contribute to the load of atmospheric dust. Those that make it through the atmosphere can cause havoc, as numerous examples of meteorites crashing into houses can attest.

Fig. 7-14. A meteorite recovered from Antarctica that is thought to be of possible Martian origin.

(COURTESY OF U.S. DEPARTMENT OF ENERGY)

Fig. 7-15. A nuclear detonation below the Earth's surface.

When a large meteorite slams into Earth, it kicks up a massive amount of sediment similar to a shallow underground nuclear detonation (FIG 7-15). The finest material is lofted high into the atmosphere, while the coarse debris falls back around the perimeter of the crater and forms a high, steep-banked rim. Not only are rocks shattered in the vicinity of the impact, but the shock waves can also cause shock metamorphism of the surrounding rocks, which changes their composition and crystal structure. The force of the impact also fuses sediment into small glassy spheres called spherules and tektites.

In South Africa, researchers found extensive deposits of these spherules. In places they were over a foot thick and dated as far back as 3.5 billion years to the Archaen eon. Spherules of Archaen-age also have been found in Australia. The spherules resemble chondrules, found on carbonaceous chondrites and in lunar soils. These discoveries might support the idea that massive meteor bombardments during the Archaen eon played a major role in shaping the surface of the planet and giving it the necessary ingredients for life.

A further consequence of large meteor impacts are shocked quartz grains, which are characterized by prominent striations across the crystal faces. Minerals such as quartz and feldspar develop these features when high-pressure shock waves exert shearing forces on their crystals and produce parallel fracture planes called lamellae. Sediments that date back 65 million years ago contain shocked quartz, common soot, and unique concentrations of irridium, a rare isotope of platinum that is relatively abundant on meteorites and comets. Sediments found throughout the world also contain the mineral stishovite, a dense form of silica that is not found anywhere on Earth except at known impact sites.

This evidence is used to support the theory that the Earth was struck by a large asteroid or comet at the end of the Cretaceous period. The impact injected huge amounts of debris into the upper atmosphere and set global-wide forest fires that sent large quantities of soot into the lower atmosphere. The combination of the two events reduced surface temperatures by several degrees for several months. The gigantic collision was so devastating that it killed the dinosaurs and three-quarters of all other known species. The collision also might have changed the Earth's orbit slightly, which would help explain why the planet has been steadily cooling ever since.

8

Mass Extinctions

CERTAIN extinction events coincide with episodes of glaciation. The living space of warmth-loving species was dramatically reduced to areas narrowly confined to the tropics during periods of global cooling. This in turn reduced the number of separate species. Species unable to migrate or adapt to colder conditions were usually the ones hardest hit. This is the main reason why major extinctions generally follow periods of climatic cooling.

Another reason for extinctions is that lowered temperatures slow down the rate of chemical reactions. Biological activity during a major glacial event functions at a lower energy state, which in turn can affect species diversity. Cooler temperatures can adversely affect plant life as well. The dinosaurs became extinct at the end of the Cretaceaus period apparently because the radical change in climate affected certain plants that were their main food source.

Cooler temperatures, however, also saw the development of new species that were better adapted to the colder conditions. It is even speculated that our own ancestors, which developed during the last ice age, were one of these new species. Mammals, which supplied their own body heat, were particularly well adapted to the colder conditions and rapidly filled all available space vacated by the dinosaurs when they became extinct. Other animals such as birds and aquatic species developed migratory habits that allowed them to escape the frigid weather for warmer climates.

MAJOR EXTINCTION EVENTS

Nineteenth century geologists recognized natural breaks in the geologic record. They developed a geologic time scale (TABLE 2-2) that was based on the appearance

TABLE 8-1. Radiation and Extinction for Major Organisms

ORGANISM	RADIATION	EXTINCTION
Marine invertebrates	Lower Paleozoic	Permian
Foraminiferans	Silurian	Permian & Triassic
Graptolites	Ordovician	Silurian & Devonian
Brachiopods	Ordovician	Devonian & Carboniferous
Nautiloids	Ordovician	Mississippian
Ammonoids	Devonian	Upper Cretaceous
Trilobites	Cambrian	Carboniferous & Permian
Crinoids	Ordovician	Upper Permian
Fishes	Devonian	Pennsylvanian
Land plants	Devonian	Permian
Insects	Upper Paleozoic	
Amphibians	Pennsylvania	Permian-Triassic
Reptiles	Permian	Upper Cretaceous
Mammals	Paleocene	Pleistocene

and disappearance of species. The geologic time scale is divided into four major eras: the Precambrian era—age of early life, the Paleozoic era—age of ancient life, the Mesozoic era—age of middle life, and the Cenozoic era—age of recent life. Except for the Precambrian era, which is poorly represented by the fossil record of past life, the eras are further divided into periods. The periods are in turn subdivided into epochs. Each time unit is distinguished from the next by the existing type of life forms. The boundaries between time units depict either a rapid expansion or a mass extinction of species.

Such catastrophes seem to conflict with uniformitarian concepts that assert that geological processes change very slowly and at steady rates. During the Phanerozoic eon, from about 600 million years ago (mya) to the present, there have been five major mass extinctions: the Ordovician (440 mya), the Devonian (365 mya), the Permian (240 mya), the Triassic (210 mya), and the Cretaceous (65 mya). (See TABLE 8-6.) In addition to these, there have been five minor extinctions. All extinction events seem to indicate biological systems were in extreme stress brought on by a change in climate or habitat.

The strongest extinction events coincide with the boundaries between major geologic time periods. The episodes of extinction appeared to be periodic, occurring roughly every 24 to 32 million years. Longer-period cycles of 80 to 90 million years appear between major mass extinctions with exceptionally strong mass extinctions occurring every 225 to 275 million years. These long-period mass extinctions might be related to the period of the Sun's revolution around the center of the Galaxy. Should the Solar System encounter a nebula or a dust cloud on its journey around the Galaxy, the dust could affect the Sun's output or block solar energy from reaching the Earth and cause climatic cooling.

The largest extinction event took place during the late Permian period, 240 million years ago (FIG. 8-1). During this time half the families, comprising 95 percent of all species, disappeared. The Permian period followed one of the greatest ice ages in geologic history, during which much of the world was blanketed by a sheet of ice. Another classic extinction event was the death of the dinosaurs (FIG. 8-2) and 70 percent of all other known species at the end of the Cretaceous period 65 million years ago. The dinosaurs, who are generally thought to have been cold-blooded, probably could not migrate from or adapt to the frigid conditions that appeared to advance rather suddenly by geologic standards.

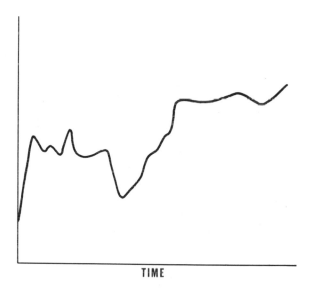

Fig. 8-1. The number of families from the Cambrian to the present. The large dip is in response to the great Permian extinction event 240 million years ago.

TIME

The periodic occurrence of lesser extinctions, approximately every 26 million years, would seem to imply a common cause for extinction events. The reason species die out on such a grand scale probably has much to do with a drastic change in the climate or sea level.

Plate tectonics, which began to spread the present continents apart 180 million years ago, could have caused changes in the climate by shifting atmospheric and oceanic circulation. This in turn could have inflicted mass extinction on a global scale. Plate tectonics, however, operates only over very long time periods with cycles of about 500 million years.

The end of the Cretaceous period was marked by a significant drop in global sea levels that occurred rather suddenly in the geologic record. The lower sea levels would have reduced the shallow water habitat area, while at the same time increasing the continental area and amount of runoff. This would have had an adverse affect on marine life and could explain the extinction of so many marine species. A fall in sea level would also have caused wider seasonal temperature variations on the continents, thereby increasing environmental stress on the dinosaurs.

One way in which the sea could have been rapidly lowered was through the growth of massive glaciers in the higher latitudes. These glaciers would have substantially affected the climate by altering the circulation patterns in the atmosphere and the

Fig. 8-2. Exposure of the Jurassic Morrison formation, Uinta Mountains, Utah, where dinosaur bones are found.

ocean. There appears to be no geologic evidence, however, for polar ice of any major quantity until the Oligocene epoch about 40 million years ago. Perhaps instead, the land rose higher from an increase in tectonic activity that uplifted extensive portions of the continents and caused the sea to regress.

The most recent mass extinction occurred toward the end of the last ice age about 11,000 years ago. It was modest compared to most of the previous extinction events and unusual in that it affected a large proportion of the big land mammal species, especially those in the Americas (FIG. 8-3). Some 70 species of large mammals became extinct in the Late Pleistocene epoch compared with about 50 species that died out

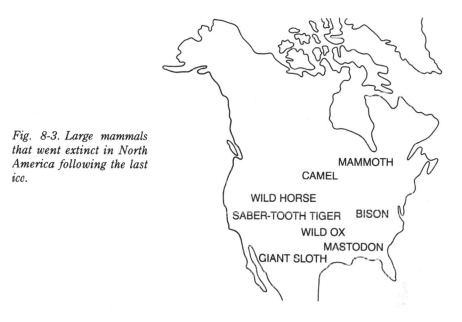

Fig. 8-3. Large mammals that went extinct in North America following the last ice.

MAMMOTH
CAMEL
WILD HORSE
SABER-TOOTH TIGER BISON
WILD OX
MASTODON
GIANT SLOTH

during the previous 3 million years. Included were the giant ground sloth, the mastadon, the mammoth, and the saber tooth cat. As the glaciers retreated, there was a global environmental readjustment. The cool, equable climate of the ice age gave way to the warmer, more seasonal climate that exists today.

The rapid environmental switch from a glacial to a postglacial age caused the forests to shrink and the grasslands to expand. This disrupted the food chain of the large animals. Deprived of their nutritional resources, they became extinct. Those mammals that were specially adapted to the cold might not have been able to stand the increasing warmth. It has even been suggested that the death blow was delivered by the hand of man. By this time, humans were efficient hunters and followed the retreating glaciers far north. Humans might have decimated the slow, lumbering creatures, just as early settlers of the American West nearly eliminated the bison.

EXTRATERRESTRIAL CAUSES OF EXTINCTIONS

Astronomical phenomena might best explain the apparent periodicity of mass extinctions. The Sun, like all the stars in the Galaxy, oscillates up and down perpendicular to the galactic plane. It completes a cycle about every 60 to 80 million years, which is nearly the same length as one of the major extinction cycles and is about twice the 30-million-year cycle of extinction. The Sun crosses the midplane and reaches maximum distance twice during each cycle, or roughly every 30 million years.

The amplitude of the oscillation is thought to be approximately 100 parsecs (326 light-years) on either side of the midplane of the Galaxy. The passage of the Sun through dense clouds located at the midplane could reduce the Earth's solar intake and initiate climatic changes that could dramatically affect life on the planet. The dust cloud, however, does not appear to be dense enough to block out the Sun during each passage through the galactic midplane, a journey that can take several million years.

Presently, our Sun is near the midplane of the Galaxy and we appear to be midway between major extinction events. The most recent ones occurred approximately 38 million and 11 million years ago. The extinction episodes therefore might coincide with the Sun's approach to its maximum distance from the galactic midplane. The extinction of the dinosaurs and large numbers of other species 65 million years ago occurred at a time when the Sun's distance from the midplane was nearly at its maximum. As the Solar System reaches the upper or lower regions of the Galaxy, it might be exposed to high levels of cosmic radiation from supernovas, which could ionize the Earth's upper atmosphere and produce a sunscreen that could cool the surface.

Periods of magnetic reversals also coincide with variations in the climate. Certain magnetic reversals coincide with the extinction of species. Nevertheless, it has not been demonstrated adequately that magnetic reversals are periodic. They might, however, be associated with other periodic phenomena such as the traversal of the Solar System through the magnetic field of the galactic plane. The galactic magnetic field is so weak, however, it is doubtful it could influence the Earth's magnetic field, which is a million times stronger. Large meteorite impacts, very strong earthquakes, or intense volcanic activity have also been cited as causes of geomagnetic field reverses. At the end of the Cretaceous period, a long period of geomagnetic stability came to an abrupt end, possibly caused by one of the most massive volcanic outbursts or meteorite impacts in geologic history.

An extraterrestrial body such as a large asteroid or comet colliding with the earth could cause almost instantaneous extinctions. Meteor Crater near Winslow, Arizona (FIG. 8-4), is nearly a mile wide and about 600 feet deep. This is comparatively small

(COURTESY OF USGS)

Fig. 8-4. Meteor Crater near Winslow, Arizona.

as meteorite craters go and relatively young, only 22,000 years old. Because this crater exists in the desert, it has escaped erosion, which has erased all but the faintest signs of other great meteorite bombardments of the distant past.

A massive comet shower involving perhaps thousands of comets might help explain the disappearance of the dinosaurs and other species. The comets would shock heat the atmosphere as they streaked toward the surface, causing nitrogen and oxygen to combine to form nitric oxide (NO). When combined with water, the NO radicals could form nitric acid, which would cause almost pure acid to rain down on the planet. Nitric oxide also destroys ozone. The erosion of the ozone layer by a comet shower could leave all life defenseless against the Sun's strong ultraviolet radiation. The expected 100 million years between comet showers, however, is much greater than the roughly 30 million years that separate extinctions in the fossil record.

The Earth can expect to incur a significant asteriod impact on the average every 50 million years, or about twice the extinction cycle. Each impact would gouge out a substantially large hole in the ground and inject huge quantities of dust into the atmosphere. In addition, heat generated by the compression of the atmosphere and impact friction could set global forest fires. Heavy amounts of dust and soot would encircle the globe and linger for months. This could cool the climate and cause a massive killing of species, especially in the tropics where species are particularly sensitive to colder conditions.

One explanation for the extinction of the dinosaurs that has received much attention lately is a collsion between the Earth and an asteroid 10 kilometers (6 miles) wide. The boundary rocks between the Cretaceous and the Triassic periods contain a thin layer of mud composed of shock-impact sediments, microtektites (small glassy spheroids created during an impact), organic carbon, an unusually high level of iridium, and the mineral stishovite, which is found only at impact sites. Irridium is an isotope of platinum that is rare on Earth, but relatively abundant in asteroids and comets.

If an exceptionally large asteroid did strike the Earth, it would have sent some 500 billion tons of sediment into the atmosphere. Soot from massive forest fires that would have been set ablaze by the friction of the impact would also have clogged the skies and placed the planet in a deep freeze. This scenario is often used to describe a nuclear winter, which would follow a full-scale nuclear war. It is also argued that if a number of large volcanoes erupted simultaneously around the world, the effect would be similar to that of a large meteorite impact.

Going back through the geologic record, scientists have uncovered additional layers of mud that contain anomalous irridium concentrations that happen to coincide with other extinction episodes. The concentrations in these layers are not nearly as strong, however, as the irridium concentrations at the end of the Cretaceous period. Whatever happened at this time appears to have been a unique event in the Earth's history.

The apparent periodicity of the extinctions might be attributed to the Earth's movement through the galactic plane. The gravity disturbance might break loose Earthbound comets from the Oort Cloud, which surrounds the Solar System about a light-year away.

It has also been suggested that the Sun has a companion, brown dwarf star that passes close by the Oort Cloud about every 26 million years. Its passage could cause a large number of comets to shower down on the Earth. The expected iridium content of comets, however, is not nearly as great as that of metallic meteorites. Hot spot volcanoes, such as those in Hawaii, can also produce significant amounts of iridium from deep within the mantle.

Large asteroid impacts would have the same environmental consequences as large volcanic eruptions. The impact of a single massive asteroid or several moderate-size ones could send enough material into the atmosphere to cause darkness at noon. This condition might last for several months. Photosynthesis would be halted and near-surface phytoplankton in the ocean would be eliminated. The effects of these extinctions would move up the food chain in a domino effect, resulting in the extinction of large marine and terrestrial species as well.

The widespread death of microscopic marine plants called calcareous nannoplankton at the end of the Cretaceous period could have triggered an extreme global heat wave that might have killed off the dinosaurs and other species. The plants produced a sulfur compound that helps form clouds. Clouds in turn reflect sunlight and prevent solar radiation from reaching the surface. Instead of a climate change causing extinctions, this might have been a mass extinction that dramatically affected the climate.

This contention is supported by the fossil record, which suggests that ocean temperatures rose by 5 to 10 degrees Celsius for tens of thousands of years beyond the end of the Cretaceous period. At the same time, more than 90 percent of the calcareous nannoplankton became extinct, along with most life from the upper portions of the ocean, for almost half a million years. The plankton might have been killed by a lack of sunlight, which was needed for photosynthesis, or by acid rain, which might have followed a massive meteorite impact. The acid in the ocean might also have been strong enough to dissolve the plankton's calcareous shell.

Catastrophic extinction events appear to be virtually instantaneous in the geologic record. It seems likely that the extinction of species occurred over perhaps a million years or more. Because of erosion or nondeposition of sedimentary strata the die-outs in the geologic record only appear to be sudden.

During the last part of the Cretaceous period, the ammonites (FIG. 8-5), which

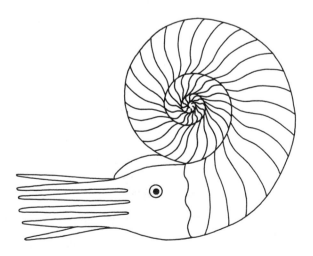

Fig. 8-5. The ammonites, which were highly successful predators, became extinct at the end of the Cretaceous period.

were a highly successful species, completely died out during an interval of about 2 million years. This was probably due to the development of more mobile and aggressive predators. Of more interest, however, is the fact that the ammonites disappeared some 100,000 years prior to the supposed asteroid impact.

TERRESTRIAL CAUSES OF EXTINCTIONS

Episodes of mass extinction also correlate reasonably well with cycles of terrestrial phenomena. The longest of these cycles is about 300 million years and is related to the cycle of convection currents in the Earth's mantle. Convection is the motion that occurs within a fluid medium as a result of a difference in temperature (FIG. 8-6).

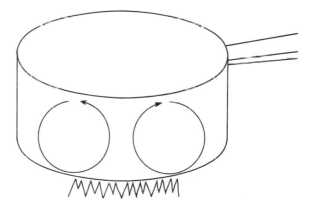

Fig. 8-6. A pot of boiling water demonstrates the principle of convection. Water heated from below rises to the surface, releases its heat, cools, and descends to the bottom again to pick up more heat.

Fluid rocks in the mantle receive heat from the core, ascend, dissipate their heat to the lithosphere, cool, and then descend back to the core to pick up more heat. The cycling of heat within the mantle is the main driving force behind all geological activity and, for that matter, all other activities taking place on the surface of the planet.

A rapid convection in the mantle led to the breakup of the supercontinents about 180 million years ago. This in turn compressed the ocean basins and caused the sea level to rise and transgress onto land. It also increased the rate of volcanic eruptions, which increased the carbon dioxide content of the atmosphere, resulting in a strong greenhouse effect that produced warm conditions worldwide. These episodes occurred from about 500 to 350 million years ago, and again from about 200 to 50 million years ago.

The second phase of the 300-million-year rock cycle was a time of low mantle convection. This caused the continents to combine into a supercontinent, the ocean basins to expand, the sea level to drop, and the sea to regress from the land. There was a reduction in atmospheric carbon dioxide because of low levels of volcanism. This produced an icehouse effect and caused colder temperatures worldwide. These conditions prevailed from about 700 to 550 million years ago, from about 400 to 250 million years ago, and during the latter part of the Cenozoic era, the period in which we are now living.

The drifting of the continents has a dramatic effect on life on Earth. Continents and ocean basins are continually being reshaped and rearranged by crustal plates that are in constant motion. When continents break up, they tend to override ocean basins. The seas are then less confined and the sea level rises several hundred feet. Low-lying, island areas are inundated by the sea. The amount of shoreline is increased dramatically, as is the area of shallow-water marine habitat. This expansion allows a larger number of species to be supported.

Mountain-building, associated with the movement of crustal plates, alters patterns of river drainages and climate, which in turn affects terrestrial habitats. When land is raised to higher elevations where temperatures are colder, especially in the higher latitudes (FIG. 8-7), glaciers grow. The location of continents in separate parts of the world can also interfere with ocean currents, which affect the distribution of the Earth's heat.

Fig. 8-7. Maclaren Glacier, Alaska.

Opposite conditions occurred when the continents were assembled into a super-continent during the latter part of the Paleozoic era. With no land to impede their travel, free-flowing ocean currents distributed heat from the tropics to the poles and kept the temperature of the planet more uniform. The ocean basins widened, which caused the sea level to drop considerably. This forced the inland seas to retreat and produced a continuous narrow continental margin around the supercontinent.

The reduced shoreline radically limited the habitat area. Unstable near-shore conditions resulted in an unreliable food supply. Species that were used to a variety of ecosystems could not cope with the limited living space and food supply. This resulted in half the terrestrial species and 95 percent of all marine species becoming extinct by the end of the Paleozoic era, 240 million years ago. This was the greatest mass extinction the planet has ever known.

Some mass extinctions also coincide with periods of glaciation. This is because temperature is perhaps the single most important factor limiting geographical distribution of species. Certain species such as coral can survive only within a narrow range of temperatures (FIG. 8-8). During warm interglacial periods, species invade all latitudes. As glaciers advance across the continents and ocean temperatures drop, however, species are forced into warmer regions with limited habitats. Species that are unable to adapt to the colder conditions, are attached to the ocean floor, or are unable to migrate are usually the ones hardest hit. The intense competition for habitat and food severely limits species diversity and thus, the number of species.

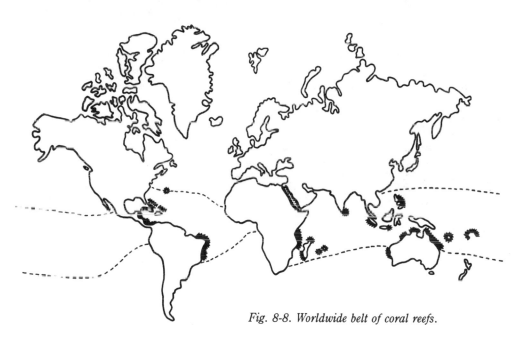

Fig. 8-8. Worldwide belt of coral reefs.

A large number of volcanic eruptions over a long period could lower global temperatures substantially by injecting huge amounts of volcanic ash and dust into the upper atmosphere. This would block out the Sun and cause mass extinctions of plants and animals. Heavy clouds of volcanic dust have a high albedo and reflect much of the solar radiation back into space (FIG 8-9). The effect would be to shade the Earth and lower global temperatures.

A 5 percent drop in the amount of solar radiation reaching the surface could cause

Fig. 8-9. The eruption cloud (left center) from the El Chichon Volcano, Mexico, viewed by satellite.

the temperature to drop by as much as 10 degrees Fahrenheit, which would be enough to bring on an ice age. The long-term cooling could cause glaciers to expand and lower the sea level, which would limit the habitat area. The lowered temperature could also adversely affect the distribution of species and confine warmth-loving species to regions in the tropics.

Another possible explanation for mass extinctions is acid rain caused by volcanic activity 100 times as intense as that occurring worldwide today. This could have caused widespread destruction of terrrestrial and marine life forms by defoliating plants and altering the pH balance of the ocean. Acid gases spewed into the atmosphere might also have depleted the ozone layer and allowed ultraviolet radiation to bath the planet. This might explain why the naked dinosaurs died out, while the small furried and feathered creatures survived.

THE AFTEREFFECTS

From the time life first began on Earth, there has always been a gradual die-out of species, a phenomenon called background extinctions. Major extinction events are separated by periods of lower extinction rates. The difference between the two is only a matter of degree. There is also a qualitative as well as quantitative distinction between background and major extinctions. Species have regularly come and gone even during optimum conditions. Those that became extinct might have lost their competitive edge and been nudged out by a superior, better adapted species.

Because a species survives extinction does not necessarily mean it is more suitable for its environment. Instead, the loser was probably developing unfavorable traits. It appears that those characteristics that permit a species to live successfully during normal periods become irrelevant when major extinction events occur. In other words, the dinosaurs might not have done anything wrong genetically. Instead, they might have been eliminated by some type of environmental disaster. The distinction between background and major extinctions might also be distorted by the fossil record, particularly when certain species are favored over others during the process of fossilization (FIG. 8-10).

Each time a mass extinction event occurs, it resets the evolutionary clock, and forces life to start anew. Those species that survive radiate outward and fill entirely new niches, from which entirely new species evolve. These new species might develop novel adaptations that give them a survival advantage over other species. The adaptations might lead to exotic species that prosper during normal times, but because of their over-specialization are incapable of surviving mass extinctions.

It appears that nature is constantly experimenting with different life forms. When one fails, it is relegated to the trash heap of extinction, never to be tried again. Once a species is gone, it is gone for good. The odds of its particular combination of genes reappearing are astronomical. This seems to place evolution on a one-way track. It perfects species to live at their optimum in their respective environments.

Had the dinosaurs escaped extinction at the end of the Cretaceous period, certain small carnivorous dinosaurs with a brain weight to body weight characteristic of early mammals could have conceivably continued to suppress the rise of the mammals. If this had happened, our own species never would have emerged as the brainiest animals on Earth.

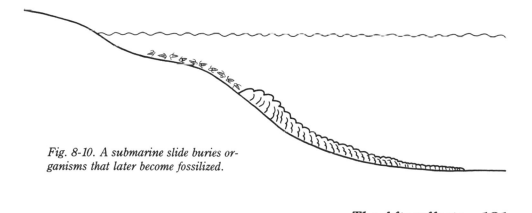

Fig. 8-10. A submarine slide buries organisms that later become fossilized.

9

The Ice Maker

EXCEPT within the last few million years, the Earth never had two permanent ice caps. In fact, it was rare when a single polar ice cap existed because of the configuration of the Earth's surface. Continents were confined to regions around the equator, which allowed warm ocean currents free access to the polar regions. This kept them ice free year round. Within the last 100 million years, however, the continents have shifted. The South Pole is now covered by a large continent, and the North Pole is covered by a nearly land-locked sea.

Most of the continental landmasses moved north of the equator and left little land in the Southern Hemisphere, which is nine-tenths ocean. This radically changed ocean current patterns, almost entirely blocking their access to the poles. Without warm ocean currents, ice formed. It will remain until the continents again shift back toward the equator, perhaps in another 50 million years.

POLAR ICE

Greenland broke away from Norway and began to separate from North America about 60 million years ago. Alaska was connected with east Siberia, which closed off the Arctic basin and isolated it from warm water currents that originated in the tropics. This resulted in the accumulation of pack ice. About 40 million years ago, Antarctica separated from Australia and drifted over the South Pole where it acquired a continent-size ice sheet. Now ice existed at both poles, a rare event in the history of the Earth. This set up a unique equator-to-pole oceanic and atmospheric circulation system.

About 3 percent of the Earth's water is locked up now in the polar ice caps. The ice caps cover about 7 percent of the Earth's surface area. The Arctic is a sea of pack-ice whose boundary is the 10 degree Celsius (50 degree Fahrenheit) July isotherm, which is the extent of the polar drift ice in summer (FIG. 9-1). The sea ice covers an average area of 4 million square miles and has an average thickness of several tens of feet. If all the pack ice in the Arctic was to melt, it would raise global

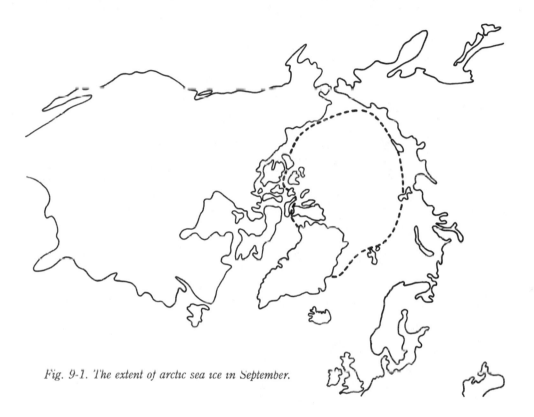

Fig. 9-1. The extent of arctic sea ice in September.

sea levels only a few feet. This increase would be enough, however, to flood seacoasts around the world. Almost 90 percent of all the ice in the world lies atop its nearly 6 million square miles, making it larger than the United States, Mexico, and Central America combined. The greatest amount of ice, by far, is in Antarctica (FIG. 9-2). Entire mountain ranges are covered by a sheet of ice 3 miles thick, with an average thickness of 1.3 miles. The total amount of ice in Antarctica is approximately 7 million cubic miles. Combined, it would make an ice cube nearly 200 miles square. The Earth's oceans currently cover an area of about 140 million square miles and have an average depth of 3 miles. Therefore, if all the ice in the Antarctic were to melt, it would raise the level of the sea by nearly 300 feet.

The additional seawater would move the shoreline inland up to 70 miles in most places (FIG. 9-3) and even more at low-lying deltas. This in turn would radically change the shape of the continents. Mississippi, Louisiana, east Texas, and major parts of Alabama and Arkansas would practically disappear. Florida, south

Fig. 9-2. Taylor Glacier region, Victoria Land, Antarctica.

Georgia, and the eastern Carolinas would also be gone. The Panama Isthmus that separates North and South America would sink out of sight. The Dutch, who worked so hard to reclaim their land from the sea, would find much of their country under water. Many islands would drown or become skeletons with only their mountainous backbones showing above the sea. Because most of the major cities of the world are either located on seacoasts or along inland waterways they would be inundated by the sea. Only the tallest skyscrapers would still poke above the water line.

THE ICE CONTINENT

Antarctica broke away from Australia and South America and drifted over the South Pole roughly 40 million years ago. About 37 million years ago, the

Fig. 9-3. Areas in Europe that would be flooded if the ice caps melted.

temperature plunged. Antarctica accumulated its first ice sheets about 2 million years later. It appears, however, that the West Antarctica ice sheet formed no earlier than about 9 million years ago. When Antarctica separated, a new ocean current was established. Easterly blowing, south polar winds circled around the South Pole and pushed against the sea as they blew across its surface. The friction between wind and wave created a permanent ocean current that completely enveloped Antarctica. It circles around the continent like a snake chasing its tail.

This isolated the frozen continent from the rest of the ocean and prevented warmer currents from reaching it. As a result, Antarctica was covered by a thick layer of ice, which dwarfed even the present ice sheet. All of the continent's land features including canyons, valleys, plains, plateaus, and mountains were covered by ice. The highest mountains were buried under at least 1,500 feet of ice. The ice sheet also extended across the sea to the tip of South America.

Despite all this ice, Antarctica is literally a desert. Its annual snowfall is only about 2 feet, which translates into about 3 inches of rain. This makes Antarctica one of the most impoverished deserts on the face of the Earth. Dry valleys that were gouged out by ice sheets (FIG. 9-4) and run between McMurdo Sound and the Transantarctic Mountains receive less than 4 inches of snowfall each year and most of that is blown away by strong winds.

The snows in Antarctica accumulate into thick ice sheets because there is virtually no melting from year to year. The mean monthly temperature at the South Pole during the summer is −28 degrees Fahrenheit; in winter it is −80 degrees. In some places, the temperature has been known to drop to −127 degrees Fahrenheit. Barren mountain peaks soar 17,000 feet into the air, and winds shriek off the ice-laden mountains and high ice plateaus at speeds of 200 miles per hour. Only automated, remote weather stations that are placed strategically across the continent (FIG. 9-5) can tolerate such frigid conditions.

Fig. 9-4. *Wright Dry Valley, Taylor Glacier region, Victoria Land, Antarctica.*

Antarctica was not always shrouded in ice. Eighty million years ago it was still attached to the southern megacontinent of gondwana. The water surrounding Antarctica was reasonably warm and there were no ice sheets. Evidence of a warm climate that supported lush vegetation and forests is found in coal seams that run through the Transantarctic Mountains and are among the most extensive coal beds in the world.

It is uncertain how many times the ice sheet on Antarctica has come and gone.

Fig. 9-5. A portable automatic weather station mounted on a sled was used to transmit weather data in Antarctica.

(COURTESY OF U.S. NAVY)

However, about 4 million years ago, great forests grew on the flanks of the Transantarctic Mountains as evidenced by discoveries of nonfossilized wood and marine fossils. In the relatively warm climate, great open seaways might have reached deep into the interior of the continent, and the central ice mass might have retreated to small interior ice sheets and high alpine glaciers. Today, only meager signs of life are found in Antarctica. They consist of blue-green algae on the bottom of small glacier-fed lakes, bacteria in the soil, and a giant wingless fly.

It is interesting to note that experiments for future excursions to Mars are performed in the dry valleys that run between McMurdo Sound and the Transantarctic Mountains. Temperatures and terrain in the region are thought to be very similar to those on the distant planet.

THE ANTARCTIC

During an antarctic expedition in the summers of 1969-70, scientists discovered a fossilized jawbone and canine tooth belonging to the mammal-like reptile lystrosaurus (FIG. 9-6) in the frigid cliffs of the Transantarctic Mountains. This unusual looking animal lived approximately 160 million years ago. It was about 2 feet long and had large down-pointing tusks. The only other known lystrosaurus fossils have been found in China, India, and southern Africa. It is highly unlikely that this fresh water reptile could swim across the salty ocean that separated the southern continents. Instead, its discovery on the frozen wastes of Antarctica was hailed as definite proof of the existence of gondwana.

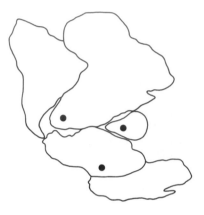

Fig. 9-6. Lystrosaurus and fossil sites on Gondwana.

In 1985, a small fossil opossum tooth, dating back 37 million years, was discovered in central Siberia. The animal could have traveled north from North America through Asia to Australia, or it could have taken a direct southern route across Antarctica (FIG. 9-7). The southern route is supported by a fossil of a South American marsupial found in Antarctica. Apparently, Antarctica was once a land bridge between the southern tip of South America and Australia, lending further support to the existence of gondwana.

The waters surrounding Antarctica are the coldest in the world (FIG. 9-8). The thermal barrier produced by the circum-Antarctic current impedes the inflow of warm currents as well as the outflow of antarctic fishes. Sea ice covers the surrounding water for at least 10 months of the year (FIG. 9-9). For 4 months each year, Antarctica is in total darkness. Even during the short summer season, the water under the ice receives less than 1 percent of the surface sunlight. The water temperature throughout the year varies from 2 to 4 degrees below freezing, but due to its high salt content the water does not freeze.

It is because of these extreme conditions that the Antarctic Sea has only about half the species diversity of the Arctic Sea. Yet, despite the fact that seawater is below freezing year round, certain species of fish thrive beneath the ice. This is because special proteins in their blood and bodily fluids act as an antifreeze-like substance that stops ice crystals from spreading through their bodies. Since all fish are cold blooded, their body temperature is essentially the same as their environment. In order to survive the long, dark Antarctic winters, these fish have adapted like no other species on Earth.

Fig. 9-7. Possible routes of marsupials from North America to Australia.

PATTERNED GROUND

One of the most barren environments on earth is the arctic tundra of North America and Eurasia. Tundra covers about 14 percent of the Earth's land surface in an irregular band that winds around the top of the world staying north of the tree line and south of the permanent ice sheets (FIG. 9-10).

Most of the ground in the arctic tundra, called permafrost, is frozen year round. Only the top few inches of the soil thaws during the short summer, even though the ground is bathed in 24-hour sunlight during the summer. This is because the thawing ice absorbs heat from its surroundings, just like the ice in a home ice cream maker. The arctic tundra is one of the most fragile environments on Earth. Small disturbances can cause a great deal of damage. Arctic haze, which originates mostly in areas of Europe and northwest Asia, tends to block out sunlight that is at a premium at this latitude.

In the permafrost regions of the Arctic, soil and rocks are fashioned into strikingly beautiful and orderly patterns that have confronted geologists for centuries. As the ground begins to thaw every summer, the retreating snows unveil a bizarre assortment of rocks arranged in a honeycomb-like network. This gives the landscape the appearance of a tiled floor (FIG. 9-11). These patterns are found in most of the northern lands and alpine regions where the soil is exposed to moisture and seasonal freezing and thawing. Polygons composed of small pebbles are a few inches across. When large boulders form protective rings around mounds of soil, the polygons can be several tens of feet in diameter.

The polygons were probably produced by processes similar to those that cause

Fig. 9-8. Coast Guard icebreaker Glacier *slices through the icy seas of Antarctica during operation Deep Freeze.*

frost heaving, which pushes rocks up through the soil. The boulders move through the soil when they are pulled from above or pushed from below. If the top of the rock freezes first, it is pulled up by the expanding frozen soil. When the soil thaws, the rock settles at a slightly higher level. The expanding frozen soil below can also heave

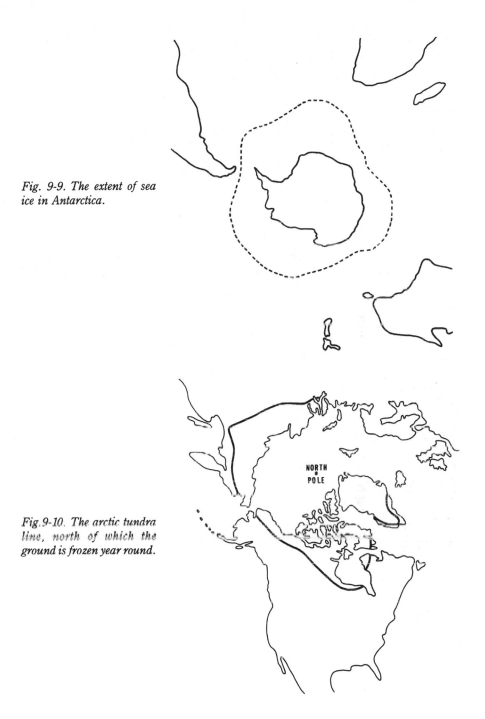

Fig. 9-9. The extent of sea ice in Antarctica.

Fig. 9-10. The arctic tundra line, north of which the ground is frozen year round.

the rock upward. After several frost-thaw cycles, the boulder finally rests on the surface.

The regular, polygonal-patterned ground in the arctic regions might have been formed by the movement of mixed composition soil upward toward the center of the

Fig. 9-11. Patterned ground on northern Alaska sea coast near Barrow.

polygonal mound and then downward under the boulders. The soil might move in convective cells (FIG. 9-12) much like the hot air in the tropics moves toward the poles. The coarser material composed of gravel and boulders is gradually shoved radially outward from the central area, while the finer materials lag behind. This idea is supported by the fact that the material in the center of the polygonal often appears to have been churned up by some giant underground mole.

THE RISING SEA

Over the last 100 years, the global sea level has apparently risen upwards of 6 inches, due mainly to the melting of the antarctic and Greenland ice sheets. At the end of the last ice age between 16,000 and 6,000 years ago, torrents of meltwater

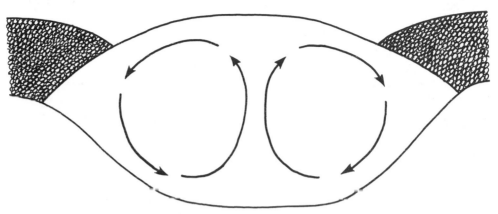

Fig. 9-12. *Soil convection is thought to be the process behind the creation of patterned ground.*

Fig. 9-13. *Damage to beach area from storm surge and high tides in Virginia Beach, Virginia.*

entered the ocean. This raised the sea level on a yearly basis only 10 times more than it is rising today. If present trends continue, the slow melting of both polar ice caps could raise the level of the oceans upwards of 12 feet by the end of the next century. This would drown much of the world's coastal plains and flood coastal cities. Even now, barrier islands and sea coasts are eroding at alarming rates (FIG. 9-13).

The Rising Sea   133

During the previous interglacial age, which ended about 127,000 years ago, the climate was actually warmer than it is today. The melting of the ice caps caused the sea level to rise about 60 feet higher than it is today (Table 9-1). If average global temperatures continue to rise, this interglacial could become as warm as the last. The warmer climate could destabilize the West Antarctic ice sheet and cause it to surge into the ocean. This rapid flow of ice could raise the sea level 15 feet, inundate the continents up to 3 miles inland, and flood trillions of dollars worth of property.

TABLE 9-1. Major Changes in Sea Level

DATE	SEA LEVEL	HISTORICAL EVENT
2200 B.C.	low	
1600 B.C.	high	Coastal forest in Britain inundated by the sea.
1400 B.C.	low	
1200 B.C.	high	Egyptian ruler Ramses II builds first Suez canal.
500 B.C.	low	Many Greek and Phoenician ports built around this time are now under water.
200 B.C.	normal	
A.D. 100	high	Port constructed well inland of present-day Haifa, Israel.
A.D. 200	normal	
A.D. 400	high	
A.D. 600	low	Port of Ravenna, Italy becomes landlocked.
		Venice is built and is presently being inundated by the Adriatic Sea.
A.D. 800	high	
A.D. 1200	low	Europeans exploit lowlying salt marches.
A.D. 1400	high	Extensive flooding in low countries along the North Sea. The Dutch begin building dikes.

At the height of the last ice age, the sea was about 400 feet below its present level. An estimated 10 million cubic miles of water were incorporated into the continental ice sheets, which covered about a third of the land surface and were three times their present volume. The coastline of the eastern seaboard of the United States was about halfway to the edge of the continental shelf, which extends eastward more than 60 miles.

On the continental shelf off the eastern United States, a step on the ocean floor has been traced for 185 miles. It might represent what was once a coastline cliff that is now totally submerged. Submarine canyons carved into bedrock 200 feet below sea level can be traced to rivers on land. These submerged valleys were carved by rivers that emptied into the sea when the water level was lowered during the last ice age.

GLACIAL SURGES

The Transantarctic Mountains divide the ice continent of Antarctica into a large East Antarctica ice mass and a smaller West Antarctica lobe, which is about the size of Greenland. Like most other continents, Antarctica has major mountain chains and deep canyons, only they are buried under a thick sheet of ice. Large flat areas beneath the ice are thought to be subglacial lakes that are kept from freezing by the interior heat of the Earth. The temperature 6,000 feet below the surface of an ice sheet can be 25 degrees Fahrenheit warmer than it is on top. The high pressures that exist at such depths can also keep water a liquid even when it is several degrees colder than its normal freezing point.

West Antarctica is traversed by rivers of solid ice that are several miles broad and flow down mountain valleys to the sea. The banks as well as the interior portions of the ice streams are marked by deep cravasses. On the bottom, muddy pools of melted water lubricate the ice streams and allow them to glide smoothly along the valley floors. These ice streams emtpy onto the great ice shelves of the Ross and Weddell seas (FIG. 9-14), where they rest below the waterline. The ice shelves are

Fig. 9-14. Antarctic ice shelves—shown in stippled areas.

surrounded by floating ice that is pinned in by small, isolated islands that are buried under the ice. This creates an unstable ice mass that may account for the number and size of the icebergs that have broken off of the Ross Ice Shelf.

The largest iceberg ever found broke off the Ross Ice Shelf in late 1987. It measured nearly 100 miles long, 25 miles wide (an area about twice the size of Rhode Island), and 750 feet thick. Antarctica discharges over a trillion tons of ice into the surrounding seas. The number of extremely large icebergs has dramatically increased over the last couple of years. There is no explanation for why so many large pieces of ice are breaking off, but it might be related to an apparent warming trend in global temperatures.

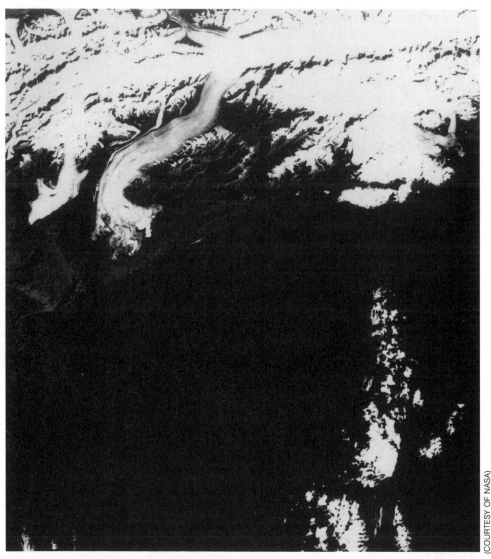

Fig. 9-15. Satellite view of the Bering Glacier in central Alaska, the largest surge glacier on the North American continent.

The flow of ice into the Southern Ocean normally compensates for the accumulation of snow in the interior. This balances the ice sheet so that its size does not change significantly. If the ice sheet grew, it would tend to cool the surrounding atmosphere and reduce the amount of snowfall. In contrast, an increased load of ice would depress the continent, which would lower the ice-sheet's surface and raise its temperature. In this way the ice is able to adjust to minor climatic changes.

Much of the ice on Antarctica accumulated during the last ice age, when the ice sheets grew 10 percent larger than they are today. The ice is not uniform, but is pervaded by internal layers. Normally ice will shatter when placed under great stress. But because of its large size a glacier acts like a viscous solid that can flow like thick molasses and creep over the landscape.

During a glacial surge, water spreads out beneath the glacier and acts like a lubricant. This allows large parts of the ice sheet to surge along the ice streams toward the sea at speeds several times faster than normal. This rapid flow of ice into the sea could raise sea levels 15 feet or more and inundate coastal areas worldwide. If enough icebergs cool the ocean surface waters and the drift ice reflects more sunlight out to space, global temperatures could cool enough to bring on a new ice age.

Today there are over 100 surge glaciers in North America (FIG. 9-15). During most of its life, a surge glacier behaves normally, moving along at a snail's pace of perhaps a couple inches a day. However, at regular intervals of 10 to 100 years, the glaciers gallop forward up to 100 times faster than normal. A surge often progresses along a glacier like a great wave, proceeding from one section to another. There is no good explanation for why glaciers surge, although they might be influenced by the climate, volcanic heat, and earthquakes. Surge glaciers, however, exist in the same areas as normal glaciers, often almost side by side. In addition, the great 1964 Alaskan earthquake failed to produce more surges than there were before (FIG 9-16).

Fig. 9-16. Avalanche on Sherman Glacier, Cordova district, Alaska, caused by the August 24, 1964 Alaskan earthquake.

(PHOTO BY A. POST, COURTESY OF USGS)

The last surge by Alaska's Hubbard Glacier (FIG. 9-17) took place around the turn of the century. For 85 years, Hubbard Glacier has been flowing toward the Gulf of Alaska at a steady rate of about 200 feet per year. Then in June 1986, the 80-mile-long river of ice surged ahead as much as 46 feet in a single day. There are 20 similar glaciers around the Gulf of Alaska that are headed toward the sea. If enough of these reach the ocean and raise global sea levels, West Antarctic ice shelves could rise and be set adrift. This would eventually raise the level of the sea even higher, which in turn would release more ice and set in motion a vicious cycle.

Fig. 9-17. Hubbard Glacier, Yakutat district, Alaska.

A total collapse of the West Antarctic ice sheet might come about as the result of a warmer climate generated by the greenhouse effect. A flood of ice would then come crashing into the Antarctic Sea. With the rise in sea level, more ice would be set free. The increased area of ice would increase the albedo, which could substantially cool the climate and cause instabilities in atmospheric and oceanic circulation systems. As a result, the Earth could be headed toward a new ice age within a person's lifetime.

10

The Greenhouse Effect

ABOUT 20 percent of the world's human population, which live in the rich industrialized nations, consume some 80 percent of its resources. They are also responsible for most of the global pollution and greenhouse warming. While people in these nations are enjoying the fruits of their modern life-style, the poorer nations are rushing to develop their share. This will dramatically increase the release of greenhouse gases and other pollutants into the atmosphere and cause global temperatures to rise considerably (FIG. 10-1) By the middle of the next century, average global temperatures will be from 5 to 10 degrees Fahrenheit warmer than they are today.

Such an increase in global temperatures would substantially change the face of the planet. Deserts could take the place of harvested rain forests. The central portions of the continents, which normally experience occasional droughts (FIG. 10-2), could become permanently dry wastelands. Once productive cropland could lose its topsoil and become a man-made desert. At the other extreme, the southern tropics could double their rainfall, causing severe flooding and ecological disaster. Regions in the colder, higher latitudes like Canada and the Soviet Union might actually benefit from increased global warming. However, much of this land would be unsuitable for agriculture and therefore could not be expected to make up for the world's rapidly dwindling agricultural lands.

Most countries will feel the adverse effects of rising sea levels. If the present melting continues, the sea could rise 6 feet by the middle of the next century. Large tracks of coastal land would simply disappear along with shallow barrier islands and coral reefs. Low-lying fertile deltas that support millions of people would drown. Delicate wetlands, where many species of marine life hatch their young, would be reclaimed by the sea. Vulnerable coastal cities would have to move farther inland or

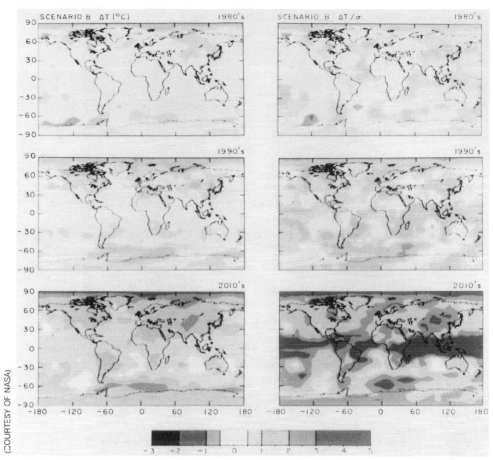

Fig. 10-1. Mean global temperature changes for the decades of the 1980s, 1990s, and 2010s.

(COURTESY OF NASA)

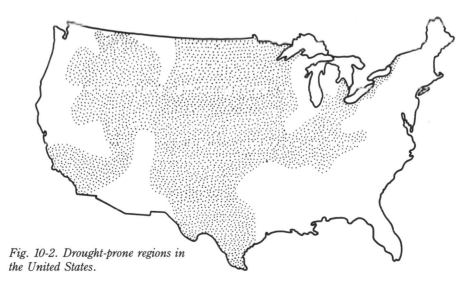

Fig. 10-2. Drought-prone regions in the United States.

build protective walls against an angry sea. Hurricanes would prowl the seas in greater numbers and at higher intensities. The effects of global warming could last for centuries, making this not the best of possible worlds.

CLIMATIC CHANGE

Ice cores taken from glaciers in central China indicate that temperatures in that region were 2 to 5 degrees Fahrenheit higher during the last 50 years than during the 50 years previous. Within 50 to 100 years, the Earth will be hotter than it has been in the past million years due to high atmospheric carbon dioxide levels, the warming will be greatest at the higher latitudes of the Northern Hemisphere. The average temperature is expected to increase 2 to 5 degrees Celsius (4 to 9 degrees Fahrenheit) with temperatures in some regions rising as much as 10 degrees Celsius. There could also be a drop of as much as 3 degrees Celsius in others.

Forests and other ecosystems could move north as much as 500 miles, bringing with them a complete reorganization of biological communities. The warming would also cause many species to become extinct. This would in turn diminish biological diversity and have an adverse effect on humans. Undesirables such as weeds and pests would over run much of the landscape. Coastal species would be especially hard hit by rising sea levels and salt water intrusion. The Earth would thus become a totally new biological world.

There is no longer any question about when greenhouse warming will come. Atmospheric scientists have little doubt that it is here already. The only question left is how to work out the intricate details of the greenhouse effect. The increase in greenhouse gases, principally carbon dioxide, methane, nitrous oxide, chlorofluorocarbon, and ozone, is blamed principally on human activities, which either directly or indirectly alter the biosphere.

The Earth has faced climate swings before, but what is different about this one is its unprecedented speed. The present warming is 10 to 40 times faster than the average rate of warming after the last ice age. In the past, plants and animals responded to changes in climate by migrating to more hospitable regions. There is even fossil evidence showing that hadrasaurs (FIG. 10-3), which were duck-billed dinosaurs that lived in Alaska during the warm Cretaceous period, migrated south when the climate grew colder. It is unlikely, however, that species can keep pace with today's rapid climate changes. Even those that might be able to would be blocked by natural or man-made barriers.

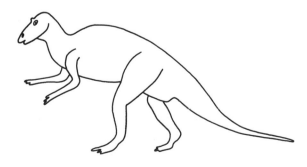

Fig. 10-3. Fossil bones of Hadrasaur found in Alaska suggest the dinosaur might have migrated south when the climate grew cold.

Some areas, particulary in the Northern Hemisphere, will dry out due to the high temperatures. This will increase the massive forest fires. Evaporation rates will increase, changing circulation patterns and actually increasing rainfall in certain areas.

Clouds are a wild card in the climate puzzle, however. There is much speculation that the added number of clouds could actually block out sunlight and cool the Earth. Presently, clouds cover half the planet and reflect 30 percent of the sunlight back into space. Clouds also block outgoing infrared radiation from escaping into space. Venus is totally shrouded by clouds that reflect 80 percent of the sunlight, yet due to a run-away greenhouse effect its average surface temperature is more than 15 times that of Earth.

Many of the clues about what physical changes will be brought on by global warming are found in the geologic record, which shows a far more complex response to climate change than previously had been expected. By the end of the Pleistocene epoch, between 14,000 and 10,000 years ago, the Earth warmed up perhaps 3 to 5 degrees Celsius. Although this temperature increase is similar to the predicted increase from the greenhouse effect, the major difference is that it was spread over a period of several thousand years and not compressed into less than a century (FIG. 10-4).

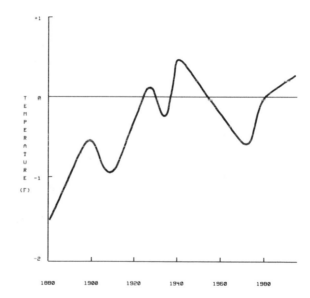

Fig. 10-4. Temperature increases over the past century.

FOSSIL FUEL COMBUSTION

In the 200 years since the Industrial Revolution, human activities have caused the level of atmospheric carbon dioxide to rise by 25 percent from 0.027 percent to 0.035 percent. The long-term increase is the result of an accelerated release of carbon dioxide from the combustion of fossil fuels. About 5.5 billion tons of carbon enter the atmosphere each year. That is about 1 ton of carbon for every person on Earth. Americans alone release some 6 tons of carbon per person per year and are responsible for up to 4 times more carbon entering the atmosphere than other industrial nations.

The destruction of the world's forests and changes in land use send an additional 2.5 billion tons of carbon into the atmosphere annually. Since the atmosphere holds about 700 billion tons of carbon, human activity alone accounts for a 1 percent annual increase.

Since the 1970s, the United States has increased its consumption of coal by about 70 percent (FIG. 10-5). In order to keep up with demand, the U. S. will have to mine about 60 percent more coal by the turn of the century. Presently, the total world coal production is about 5 billion tons annually. The U. S. accounts for about half of the coal mined and consumed by the free world.

Fig. 10-5. Huge coal-fired generating plant near Muscogee, Oklahoma.

The consumption of fossil fuels will continue to increase well into the next century. When stocks of petroleum begin to dwindle, they mostly will be replaced by coal, which is a much dirtier fuel. It releases nearly twice as much carbon into the atmosphere per unit of energy as petroleum. Underdeveloped nations that want to industrialize in order to improve their standard of living will play a major role in the increased consumption of fossil fuels.

The Earth's average annual temperature has warmed by at least 0.3 to 0.4 degrees Celsius over the last century. The surface warmed significantly during the early part of this century, cooled slightly from 1940 to 1972, and then began to warm sharply. The oceans are undergoing a gradual but significant warming of from 0.05 to 0.1 degrees Celsius per year. Since 1973, measurements taken from satellites have revealed that polar sea ice has shrunk by 6 percent. The oceans can store large amounts of heat, however, and respond sluggishly to rapid temperature increases on land. Sea temperatures could take 1000 years or more to react fully to a doubling of carbon dioxide levels.

The capacity of the ocean to absorb carbon dioxide is almost limitless. Carbon dioxide moves from the atmosphere through the mixed layer of the ocean and into the oceanic depths very slowly and at a nearly constant rate. This rate is only half the amount of carbon dioxide currently being released into the atmosphere by the combustion of fossil fuels. The problem is exacerbated by the fact that all living matter is also a net source of atmospheric carbon dioxide, which is equal to the combustion of fossil fuels. In addition, volcanic eruptions produce substantial amounts of carbon dioxide. Without man's contribution, the Earth would be in equilibrium with the same amount of carbon dioxide being absorbed by the ocean as that being naturally produced.

An increase in surface temperature brought on by a doubling of the amount of atmospheric carbon dioxide could have a global effect on precipitation patterns (FIG. 10-6). Areas between 20 and 50 degrees north latitude and 10 to 30 degrees south latitude would experience a marked decrease in precipitation, which would bring about desert conditions. These changes would have a profound effect on the distri-

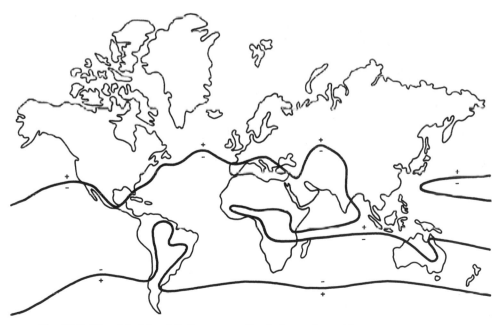

Fig. 10-6. The global precipitation-evaporation balance. In positive areas, precipitation exceeds evaporation. In negative areas, evaporation exceeds precipitation.

bution of water resources that would be desperately needed for irrigation of agricultural lands. Not only would rainfall be diminished, but the higher temperatures would increase evaporation. River flow could decline by 50 percent or more with some rivers drying up entirely. Major groundwater systems would also be adversely affected and water wells could go dry.

Melting ice caps, which could substantially raise sea levels and inundate coastal regions and agriculturally rich deltas would be another consequence of global warming. Higher sea levels would also alter the shapes of the continents and drown low-lying barrier islands and atolls. For every 1 foot increase in the level of the sea, 100 to 1,000 feet of shoreline would disappear. Therefore, global warming is of no particular benefit to mankind.

DEFORESTATION

Tropical rain forests, particularly those in the Amazon Basin of South America (FIG. 10-7), are being destroyed at a rate of about 50,000 square miles, or an area the size of the state of New York, every year. If the present global deforestation continues, the world's rain forests will be all but obliterated by the early part of the next century. The forests are disappearing through small-scale slash and burn agriculture and large-scale timber harvesting methods. Some tropical forest fires are so large that gigantic smoke plumes viewed from space rise up like great cumulus clouds (FIG. 10-8).

Fig. 10-7. The amazon rain forest of South America.

Fig. 10-8. Smoke from forest fires in the Amazon River basin completely obscures the ground as seen from the space shuttle Discovery in early 1989.

Rain forests in Hawaii are cut down and burned in huge electrical generating plants. Unwanted trees and brush are burned and the bare soil, which is now totally denuded of all vegation, is left unattended. When the rains come, the denuded soil is washed away by fierce flash floods, leaving mostly bare rock behind. Without the soil, the chances of recovery are next to none, and the forests for all practical purposes are gone for good.

The forests of the world are extensive and act as natural climate controllers. Desertification brought on by the loss of the forests increases surface albedo, which reflects sunlight into space. This loss of solar energy could change precipitation patterns and result in decreased rainfall, particularly in the rain forests, which are currently responsible for half their own rainfall. This could cause further stress and make trees more susceptible to disease. Dry conditions also promote forest fires that produce soot, which absorbs solar radiation. This heats the atmosphere, and produces temperature imbalances that can send abnormal weather to all parts of the world.

Forests also store a great deal of carbon. The clearing of forested land for agriculture, especially in the tropics, releases large amounts of carbon into the atmosphere. Although there is a net accumulation of carbon in the forests of North America and Europe, it is insignificant compared with the losses in the tropical regions. The increased carbon dioxide content heightens the greenhouse effect, which can substantially change global weather patterns as the world heats up.

Deforestation also poses a significant threat to the ozone layer, which protects the Earth from the Sun's harmful ultraviolet rays. Every spring since the beginning of the 1980s, a vast ozone hole has opened over Antarctica (FIG. 10-9). The ozone depletion is thought to be caused mainly by a chemical reaction that breaks apart ozone molecules. Chief among the ozone-destroying chemicals are chlorofluorocarbons and nitrous oxides. Chlorofluorocarbons are used as refrigerants, propellants, and solvents in the manufacture of foam plastics. Fossil fuel combustion and clear-cutting of timber release tremendous amounts of nitrous oxide into the atmosphere.

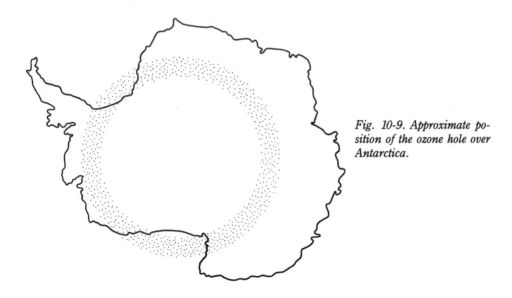

Fig. 10-9. Approximate position of the ozone hole over Antarctica.

Living things and dead organic matter in the soil hold several times the amount of carbon that is in the atmosphere. The harvest of forests, the extension of agriculture, and the destruction of wetlands speed up the decay of humus, which turns into carbon dioxide and enters the atmosphere. Agricultural lands do not store as much carbon as the forests they replace.

The world's forests conduct more photosynthesis than any other type of vegetation. They incorporate from 10 to 20 times more carbon per unit area than does cropland or pastureland. The great forests have a pronounced seasonal influence on the carbon dioxide content of the atmosphere. The seasonal variation in the atmospheric concentration of carbon dioxide (FIG. 10-10) is correlated with the pulse of pho-

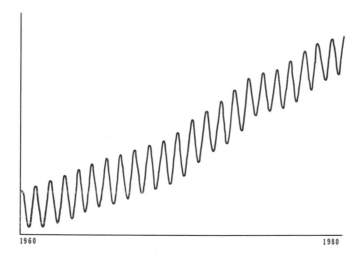

Fig. 10-10. Seasonal variation and long-term increase in atmospheric carbon dioxide concentrations.

1960   1980

tosynthesis in the forests of the middle latitudes that occurs during the summer. The change is substantially less in the Southern Hemisphere because the smaller land-mass limits the forests area.

Forests also have the potential to store carbon in quantities that are sufficiently large enough to affect the carbon dioxide content in the atmosphere. As the stores of carbon in the trees are being released into the atmosphere, the reduction of the forests is weakening the ability of the trees to remove excess atmospheric carbon dioxide and replenish the atmosphere with oxygen.

One way of slowing down the carbon dioxide buildup in the atmosphere is to plant more trees. By doubling the volume of forest growth each year, the major fossil-fuel burning nations could delay the onset of the greenhouse effect for perhaps a decade or more. The destruction of the rain forests, however, would have to be discontinued. It would take a forest covering nearly 3 million square miles, or about the size of Australia, to fully restore the Earth's carbon dioxide balance. This would be an area equal to all the forested land that has been cleared since the advent of agriculture 10,000 years ago.

EXTENSION OF AGRICULTURE

The rain forests comprise about 6 percent of the land surface. Yet they are home to more than two-thirds of all terrestrial species, including 80,000 plant species and possibly 30 million animal species—mostly insects. Present extinction rates appear to be running at about 100 species per day and are expected to escalate every year. If these trends continue, half the world's species could become extinct sometime during the next century.

As developing nations attempt to raise their standard of living, one of the first steps they take is to clear the forests and drain the wetlands for agriculture. Much of the forested land is cleared by primitive slash-and-burn methods. Trees are set afire and their ashes are used to fertilize the thin, nutrient-poor soil. In impoverished countries without other fuels, trees are also cut down to provide firewood for cooking and winter heat.

The world's wetlands are perhaps the richest of all ecosystems and produce upwards of eight times as much plant matter per acre as an average wheat field. A large variety of plants and animals make wetlands their home. Coastal wetlands support valuable fisheries. About two-thirds of the shellfish in U. S. waters rely on these areas for spawning and nursery grounds. Wetlands act as natural filters, removing sediments and some types of water pollution. They also protect coasts against storms and erosion (FIG. 10-11).

Fig. 10-11. Beach wave erosion at Grand Isle, Louisiana.

Despite their importance to near shore aquatic life, wetlands are being destroyed for agricultural land at an alarming rate. Almost 90 percent of recent wetland losses in the United States, about 1,000 acres per day, have been for agricultural purposes. The urgent need to feed the hungry is a major reason for destroying the wetlands of the Third World. The cost is the destruction of local fisheries and breeding grounds for marine species and wildlife. Like the rain forests, the destruction of the wetlands is irreversible. Their loss might cause the elimination of more species.

THE COMING OF ICE

Geological evidence indicates that major glaciers traveled over much of the land surface at various times during the Earth's history. Most of the evidence for massive glaciation comes from deposits of glacial rocks called moraines and tillites, which are found and dated fairly accurately in many parts of the world. Analysis of deep-sea sediments and glacial cores provide accurate information about some of the events that took place during the most recent ice age.

In order for the ice sheets to have formed, the temperature had to have gotten cooler, but not necessarily a whole lot cooler than it is today. (Extremely cold temperatures would also inhibit the precipitation of snow). Once the ice was established, it became self-perpetuating. The albedo effect of light colored ice reflects most of the sunlight back out into space so very little is left to heat the snowpack.

One of the hazards of greenhouse warming is the prospect of a glacial surge (FIG. 10-12). As the present climate continues to warm, the antarctic ice sheets could become unstable. This would cause them to calve off into the ocean and make additional sea ice. An increased area of ice in the Southern Ocean could form a gigantic ice shelf, which could cover as much as 10 billion square miles.

The additional sea ice would greatly increase the Earth's albedo and cause temperatures in the Southern Hemisphere to drop. The ice would chill the waters in the Southern Ocean and ocean currents would move the cold water northward. One such current that originates in Antarctica sends cold, salty bottom water as far north as New Jersey. The drop in ocean temperature could dramatically change the path of ocean currents and disrupt weather patterns significantly enough to bring on a new ice age.

During the coldest part of the last glacial period, 18,000 years ago, the concentration of carbon dioxide in the atmosphere was about half that of the preindustrial age and about a quarter of what it is today. As the ice sheets began to melt, about 16,000 years ago, carbon dioxide levels began to increase rapidly up until about 10,000 years ago. The excess carbon dioxide dissolved in the ocean, where microscopic organisms converted it into carbonates. The carbon dioxide then became permanently locked up in the Earth's crust.

Since the industrial era, the level of atmospheric carbon dioxide has again been rapidly increasing. An industrialization continues, more fossil fuel is burned, more forests are cut down, and the concentration of carbon dioxide and other trace gases in the atmosphere dramatically is increased. By the end of this century, these greenhouse gases might cause an appreciable warming of the Earth.

For over a million years, as many as 13 ice ages have come and gone on an almost regular basis, with about one every 100,000 years. These dates agree well with the

Fig. 10-12. The Yentna Glacier surging toward the sea from Mount McKinley National Park.

Milankovitch model of the ice ages, which relies on the Earth's orbital variations. The ice ages have been preceded by short interglacial ages that lasted about 10,000 years. The present interglacial is nearly over. Analysis of carbon dioxide levels in sea-floor sediments suggests there were higher atmospheric carbon dioxide concentrations during the preceding warm interglacial age than there are in this one. Yet, the warm climate did not halt the onset of the last ice age. Perhaps it even might have triggered it.

Furthermore, natural causes have varied considerably the atmospheric carbon dioxide levels within as little as 100 years. These sharp jumps were too rapid for the slow exchange of carbon dioxide from the atmosphere to the ocean, which takes about 1,000 years. Man's carbon dioxide contribution could be considered just another short thermal pulse that will last at most a few hundred years. On a geologic time scale, this influence would be insignificant and powerless in stopping another ice age. Therefore, ice ages will probably continue to come and go just as they always have done.

Glossary

ablation   The process of melting of a glacier.

abrasion   Erosion by friction, generally caused by rock particles carried by running water, ice, or wind.

absolute zero   The temperature at which all molecular motion ceases, which is −273 degrees Celsius or 0 degrees Kelvin.

absorption   The process by which radiant energy incident on any substance is retained and converted into heat or other form of energy.

abyssal   The deep ocean, generally over a mile deep.

adsorption   The adhesion of a thin film of liquid or gas to the surface of a solid substance.

accretion   The accumulation of celestial dust by gravitational attraction into a planetesimal, asteroid, moon, or planet.

advection   The horizontal movement of air, moisture, or heat.

aerosol   A mass made of solid or liquid particles dispersed in air.

albedo   The amount of sunlight reflected from an object.

alluvium   Sediment deposited in a stream.

alpine glacier   A mountain glacier or a glacier in a mountain valley.

amplitude   The length of back-and-forth motions.

Andromeda Galaxy   The nearest large spiral galaxy 2.3 million light years away.

angular momentum   a measure of an object or orbiting body to continue spinning.

anticline   Folded sediments that slope downward away from a central axis.

aphelion   The point at which the orbit of a planet is at its farthest point from the Sun. In the case of the Earth, it occurs in early July.

apogee   The point at which a spacecraft is furtherest from the Earth.

aquifer   A subterranean bed of sediments through which groundwater flows.

ash, volcanic   Fine pyroclastic material injected into the atmosphere by an erupting volcano.

asteroid   A rocky or metallic body, which orbits the Sun between Mars and Jupiter.

asthenosphere   A layer of the upper mantle, between 50 and 200 miles below the surface, that is more plastic than the rock above and below and might be in convective motion.

atmospheric pressure   The weight per unit area of the total mass of air above a given point; also called barometric pressure.

aurora   Luminous bands of colored light seen near the poles due to cosmic-ray bombardment of the upper atmosphere.

axis   A straight line about which a body rotates.

basalt   A volcanic rock that is dark in color and usually quite fluid in the molten state.

biogenic   Sediments composed of the remains of plant and animal life, such as shells.

black smoker   Superheated hydrothermal water that rises to the surface at a midocean ridge. The water is supersaturated with metals. When it exits through the sea floor, the water quickly cools and the dissolved metals precipitate, resulting in black, smoke-like effluent.

boulder   A large rock fragment.

caldera   A large pit-like depression found at the summits of some volcanoes, which was formed by great explosive activity and collapse.

calving   Formation of icebergs when they break off of glaciers upon entering the ocean.

carbonate   A mineral containing calcium carbonate such as limestone and dolostone.

Cenozoic   An era of geologic time comprising the last 65 million years.

centrifugal force   The action by which a body is propelled away from the center of rotation.

center of mass   The point at which all the mass of a body or system of bodies might be considered for calculating the gravitational effect when a force is applied.

circulation   The flow pattern of a moving fluid.

cirque   A glacial erosion feature, which produces an amphitheater-like head in a glacial valley.

condensation   The process whereby a substance changes from the vapor phase to liquid or solid phase; the opposite of evaporation.

conduction   The transmission of energy through a medium.

conglomerate   A sedimentary rock composed of welded fine-grained and coarse-grained rock fragments.

conservation law   A law that states that a particular quantity does not change or is conserved in a given physical process.

continent   A slab of light, granitic rock that floats on the denser rocks of the upper mantle.

continental drift   The concept that the continents have been drifting across the surface of the Earth throughout geologic time.

continental glacier   An ice sheet covering a portion of a continent.

continental shelf   The offshore area of a continent in shallow sea.

continental shield   Ancient crustal rocks upon which the continents grew.

continental slope   The transition from the continental margin to the deep-sea basin.

convection   A circular, vertical flow of a fluid medium due to heating from below. As materials are heated they become less dense and rise, while cooler, heavier materials sink.

coral   Any of a large group of shallow-water, bottom-dwelling marine invertebrates that build reef colonies in warm waters.

core   The central part of a planet, which consists of a heavy iron-nickel alloy.

Coriolis effect   The apparent force that deflects wind and ocean currents and causes them to curve in relation to the rotating Earth.

correlation   The tracing of equivalent rock exposures over distance.

cosmic dust   Small meteoroids existing in dust bands possibly created by the disintegration of comets.

cosmic rays   High-energy charged particles that enter the Earth's atmosphere from outer space.

crater, meteoritic   A depression in the crust produced by the bombardment of a meteorite.

crater, volcanic   The inverted conical depression found at the summit of most volcanoes, which was formed by the explosive emission of volcanic ejecta.

craton   The stable interior of a continent that is usually composed of the oldest rocks on the continent.

Cretaceous   A period of geologic time encompassing from 135 to 65 million years ago.

crevasse   A deep fissure in the Earth or a glacier.

crust   The outer layers of a planet's or moon's rocks.

crustal plate   One of several plates comprising the Earth's surface rocks.

Currie point   The temperature at which iron molecules align to a magnetic field upon cooling.

density   The amount of any quantity per unit volume.

deuterium   A heavy hydrogen atom that contains one neutron.

diapir   The buoyant rise of a molten rock through heavier rock.

differentiation   The separation of solids or liquids according to their weight with heavy mass sinking and light material rising toward the surface.

diffusion   The exchange of fluid substance and its properties between different regions in the fluid, as a result of small, almost random motions of the fluid.

drought   A period of abnormally dry weather sufficiently prolonged for the lack of water to cause serious deleterious effects on agricultural and other biological activities.

drumlin   A hill of glacial debris that faces in the direction of glacial movement.

dune   A ridge of wind-blown sediments usually in motion.

dynamo   A device that converts rotational energy into electrical or magnetic energy.

earthquake   The sudden breaking of the Earth's rocks.

East Pacific Rise   A mid-ocean spreading center that runs north-south along the eastern side of the Pacific. The predominant location upon which hot springs and black smokers have been discovered.

ecliptic   The plane in which the Earth's orbit traces an elliptical path around the Sun.

electromagnetic radiation   The energy from the Sun that travels through the vacuum of space to reach the Earth as electromagnetic waves.

electron   Negative particles of small mass that orbit the nucleus and are equal in number to the protons.

element   A material that consists of only one type of atom.

elliptical galaxy   A galaxy whose structure is smooth and amorphous, without spiral arms.

eolian   A deposit of wind-blown sediment.

epoch   A geological time unit shorter than a period and longer than an age.

equinox   Either of the two points of intersection of the Sun's path and the plane of the Earth's equator.

erratic   A glacially deposited boulder that is far from its source.

esker   Curved ridges of glacially deposited material.

estuary   Tidal inlet along a coast.

evaporation   The transformation of a liquid into a gas.

evaporite   The deposition of salt, anhydrite, and gypsum from evaporation in an enclosed basin of stranded sea water.

evolution   The tendency of physical and biological factors to change with time.

exosphere   The outermost portion of the atmosphere that is in contact with space.

extraterrestrial   Pertaining to all phenomena outside the Earth.

extrusive Any igneous volcanic rock that is ejected onto the surface of a planet or moon.

fault A breaking of rocks caused by earth movements.

fissure A large crack in the crust through which magma might escape from a volcano.

foraminifera Calcium carbonate secreting organisms that live in the surface waters of the oceans. After death their shells form the primary constituents of limestone and sediments, which are deposited on the sea floor.

fossil Any remains, impression, or trace in rock of a plant or animal of a previous geologic age.

formation A combination of rock units that can be traced over distance.

frequency The rate at which crests of any wave pass a given point.

frost heaving The lifting of rocks to the surface by the expansion of freezing water.

frost polygons Polygonal patterns of rocks that form from repeated freezing.

galaxy A large gravitationally bound cluster of stars.

gamma rays Photons of very high energy and short wave length. They are the most penetrating of electromagnetic radiation.

general relativity The theory of gravitation developed by Albert Einstein.

glacier A mass of moving ice.

glossopteris A Late Paleozioc plant that existed on the southern continents. It is not found on the northern continents, which confirms the existence of Gondwana.

gondwana A southern supercontinent of Paleozoic time, which consisted of Africa, South America, India, Australia, and Antarctica. It broke up into the present continents during the Mesozoic era.

granite A coarse-grained, silica-rich rock that consists primarily of quartz and feldspars. It is the principal constituent of the continents and is believed to be derived from a molten state beneath the Earth's surface.

gravity A force by which bodies are attracted to one another according to their mass.

greenhouse effect The trapping of heat in the atmosphere principally by carbon dioxide.

groundwater The water derived from the atmosphere that percolates and circulates below the surface of the Earth.

guyot An undersea volcano that reached the surface of the ocean, whereupon its top was flattened by erosion. Later, subsidence caused the volcano to sink below the surface preserving its flat top appearance.

half-life   The time for one-half the atoms of a radioactive element to decay into another element.

helium   The second lightest and second most abundant element in the universe. It is composed of two protons and two neutrons.

Holocene   A geological time period covering the last 10,000 years.

horn   A pyravidal peak on a mountain formed by glacial erosion.

hot spot   A volcanic center that has no relation to plate boundary location; an anomalous magma generation site in the mantle.

hydrocarbon   A molecule consisting of carbon chains with attached hydrogen atoms.

hydrogen   The lightest and most abundant element in the universe. It is composed of one proton and one electron.

hydrosphere   The water layer at the surface of the Earth.

hydrothermal   Relating to the movement of hot water through the crust.

iceberg   A portion of a glacier that broke off upon entering the sea.

ice cap   A polar cover of ice and snow.

igneous rocks   All rocks that have solidified from a molten state.

impact   The point on the surface upon which an object lands.

inertia   Inherent resistance to applied force.

infrared   Heat radiation with a wavelength between red light and radio waves.

insolation   All solar radiation impinging on a planet.

interglacial   A warming period between glacial periods.

intrusive   Any igneous body that has solidifed in place below the surface of the Earth.

ionization   The process whereby electrons are torn off previously neutral atoms.

ionosphere   The atmospheric shell that is characterized by high ion density and extends from 40 miles to very high regions of the atmosphere.

iridium   A rare isotope of platinum that is relatively abundant on meteorites.

isostasy   The idea that the crust is buoyant and rises or sinks depending on its weight.

isotope   A variety of an element with a different number of neutrons in the nucleus.

jet stream   Relatively strong winds concentrated within a narrow belt that is usually found in the tropopause.

Kelvin   A temperature scale, similar to the Celsius scale with its zero point placed at absolute zero, or -273 degrees C.

kenetic energy   The energy that a moving body possesses as a consequence of its motion.

kettle   A depression in the ground caused by a buried block of glacial ice.

landslide   Rapid downhill movement of earth materials often triggered by earthquakes.

latent heat   Heat absorbed when a solid changes to a liquid or a liquid to a gas with no change in temperature, or heat released in the reversed transformations.

lateral moraine   The material deposited by a glacier along its sides.

Laurasia   The northern supercontinent of the Paleozoic that consisted of North America, Europe, and Asia.

lava   Molten magma after it has flowed out onto the surface.

light-year   The distance that electromagnetic radiation, principally light waves, can travel in a vacuum in one year or approximately 6 trillion miles.

limestone   A sedimentary rock that consists mostly of calcite.

lithosphere   A rigid outer layer of the mantle, typically about 60 miles thick. It is over ridden by the continental and oceanic crusts and is divided into segments called plates.

magma   A molten rock material generated within the Earth. It is the constituent of igneous rocks, including volcanic eruptions.

magnetic field reversal   A reversal of the north-south polarity of a planet's magnetic poles.

magnetosphere   The region of the Earth's upper atmosphere in which the Earth's magnetic field controls the motion of ionized particles.

magnitude   The relative brightness of a celestial body.

mantle   The part of a planet below the crust and above the core that is composed of dense iron-magnesium-rich rocks.

maria   Dark plains on the lunar surface caused by massive basalt floods.

mass   The measure of the amount of matter in a body.

mean temperature   The average of any series of temperatures observed over a period of time.

mesosphere   A region of the Earth's atmosphere between the stratosphere and thermosphere. It extends 24 to 48 miles above the Earth's surface. Also, the rigid part of the Earth's mantle below the asthenosphere and above the core.

Mesozoic   Literally the period of middle life between 230 and 65 million years ago.

meteorite   A metallic or stony body from space that enters the Earth's atmosphere and impacts on the Earth's surface.

methane   A hydrocarbon gas liberated by the decomposition of organic matter.

micron   A unit of measurement equivalent to one-thousandth of a millimeter.

mid-Atlantic ridge   The seafloor spreading ridge of volcanoes that marks the extensional edge of the North American and South American plates to the west and the Eurasion and African plates to the east.

midocean ridge   A submarine ridge along a divergent plate boundary where a new ocean floor is created by the upwelling of mantle material.

monsoon   A seasonal wind accompanying temperature changes over land and water from one season of the year to another.

moraine   A ridge of erosional debris deposited by the melting margin of a glacier.

nebula   An extended astronomical object with a cloud-like appearance. Some nebulae are galaxies; others are clouds of dust and gas within our galaxy.

neutrino   A small electrically neutral particle having weak nuclear and gravitational interactions.

neutron   A particle with no electrical charge that has roughly the same weight as the positively charged proton, both of which are found in the nucelus of an atom.

nova   A star that suddenly brightens during its final stages.

Oort cloud   The collection of comets that surround the Sun about a light-year away.

orbit   The circular or eliptical path of one body around another.

orogeny   A process of mountain building by tectonic activity.

outgassing   The loss of gas within a planet as opposed to degassing, or loss of gas from meteorites.

ozone   A molecule consisting of three atoms of oxygen that exists in the upper atmosphere and filters out ultraviolet light from the Sun.

paleomagnetism   The study of the Earth's magnetic field, including the position and polarity of the poles in the past.

paleontology   The study of ancient life forms, based on the fossil record of plants and animals.

Paleozoic   The period of ancient life between 570 and 230 million years ago.

Pangaea   An ancient supercontinent that included all the lands of the Earth.

Panthalassa   The great world ocean that surrounded Pangaea.

parallax   The difference in direction of an object as seen from two different points.

parigee   The closest point at which a spacecraft orbits the Earth.

perihelion   The point at which the orbit of a planet is at its nearest to the Sun. In the case of the Earth, it occurs in early January.

permafrost   Permanently frozen ground.

permeability   The ability to transfer fluid through cracks, pores, and interconnected spaces within a rock.

photon   A packet of electromagnetic energy, generally viewed as a particle.

photosynthesis   The process by which plants create carbohydrates from carbon dioxide, water, and sunlight.

placer   A deposit of rocks left behind from a melting glacier. Also, any ore deposit that is enriched by stream action.

planetesimals   Small celestial bodies that might have existed in the early stage of the Solar System.

plasma   A collection of positive and negative charges that are free to move independently of each other and are usually formed by stripping electrons from their nuclei.

plate tectonics   The theory that accounts for the major features of the Earth's surface in terms of the interaction of lithospheric plates.

pluvial lake   Lake formed by rain water during an ice age.

polar wandering   Movement of the geographic poles.

porosity   The percent of a rock that consists of pore spaces between crystals and grains, usually filled with water.

precession   The slow change in direction of the Earth's axis of rotation due to the gravitational action of the Moon on the Earth.

precipitation   Products of condensation that fall from clouds as rain, snow, hail, or drizzle.

proton   A large particle with a positive charge in the nucleus of an atom.

quartz   A common igneous rock-forming mineral of silicon dioxide.

radiation   The process by which energy from the Sun is propagated through a vacuum of space as electromagnetic waves. A method, along with conduction and convection, of transporting heat.

radioactivity   An atomic reaction releasing detectable radioactive particles.

radiometric dating   The determination of how long an object has existed by chemical analysis of stable versus unstable radioactive elements.

radionuclide   A radioactive element that is responsible for generating the Earth's internal heat.

reef   The biological community that lives at the edge of an island or continent. The shells form a limestone deposit that is readily preserved in the geological record.

recessional moraine   A glacial moraine deposited by a retreating glacier.

remanent magnetism   A permanently induced magnetic field in a rock.

reversed magnetism   A geomagnetic field with a reverse polarity from that of the present one.

revolution   The motion of a celestial body in its orbit, such as the Earth around the Sun.

rotation   The turning of a body about an axis.

saltation   The movement of sand grains by wind or water.

satellite   A body that orbits another body; in particular an artificial spacecraft designed for planetary exploration and communication.

sea floor spreading   The theory that the ocean floor is created by the separation of lithospheric plates along the midocean ridges, with new oceanic crust formed from mantle material that rises from the mantle to fill the rift.

seamount   A submarine volcano.

sedimentary rock   A rock composed of fragments cemented together.

seismic sea wave   An ocean wave related to an undersea earthquake.

shield   Areas of the exposed Precambrian nucleus of a continent.

shield volcano   A broad, low lying volcanic cone built up by lava flows of low viscosity.

solar flare   A short-lived bright event on the Sun's surface that causes greater ionization of the Earth's upper atmosphere due to an increase in ultraviolet light.

solar wind   An outflow of particles from the Sun that represents the expansion of the corona.

solstices   The twice yearly occurrence when the apparent distance of the Sun from the equator is at its greatest.

soluable   Refers to a substance that dissolves in water.

spiral galaxy   A galaxy, like our own, with a prominent central bulge embedded in a flat disk of gas, dust, and young stars that wind out in spiral arms from the nucleus.

storm surge   An abnormal rise of the water level along a shore as a result of wind flow in a storm.

stratosphere   The upper atmosphere above the troposphere that is about 10 miles above sea level.

striation   Scratches on bedrock made by rocks embedded in a moving glacier.

subduction zone   An area where the oceanic plate dives below a continental plate into the asthenosphere. Ocean trenches are the surface expression of a subduction zone.

sunspot   A region on the sun's surface that is cooler than surrounding regions and affect radio transmissions on Earth.

supernova   An enormous stellar explosion in which all but the inner core of a star is blown off into interstellar space. It produces as much energy in a few days as the sun does in a billion years.

surge glacier   A continental glacier that heads toward the sea at a high rate of advance.

synod   The alignment of the sun, planets, and their accompanying moons.

tectonic activity   The formation of the Earth's crust by large scale earth movements throughout geologic time.

tephra   All clastic material from dust particles to large chunks that are expelled from volcanoes during eruptions.

terrestrial   All phenomena pertaining to the Earth.

Tethys Sea   The hypothetical mid-latitude area of the oceans that separated the northern and southern continents of Gondwanaland and Laurasia some hundreds of million years ago.

thermosphere   The outermost layer of the atmosphere in which the temperature increases regularly with height.

tide   A bulge in the ocean produced by the Moon's gravitational forces on the Earth's oceans. The rotation of the Earth beneath this bulge causes the sea level to rise and fall.

till   Material deposited directly by a glacier.

transgression   A rise in sea level that causes flooding of the shallow edges of continental margins.

troposphere   The lowest 9 to 12 miles of the Earth's atmosphere, which is characterized by decreasing temperature with height.

tundra   Permanently frozen ground at high latitudes and high altitudes.

ultraviolet   Invisible light that has a wavelength shorter than visible light and longer than x-rays.

van Allen belts   regions of high-energy particles that are trapped by the Earth's magnetic field.

varve   Thinly laminated lake bed sediments.

viscosity   The resistance of a liquid to flow.

volcano   A fissure or vent in the crust through which molten rock rises to the surface to form a mountain.

volcanism   Any type of volcanic activity.

x-rays   electromagnetic radiation of high-energy wave lengths above ultraviolet and below gamma rays.

Bibliography

EARTH ORIGINS

Boss, Alan P. "Origin of the Moon." *Science* Vol. 231 (January 24, 1986): 341–345.

Cairns-Smith, A. G. "The First Organisms." *Scientific American* Vol. 252 (June 1985): 90–100.

Heppenheimer, T. A. "Journey to the Center of the Earth." *Discover* Vol. 8 (November 1987): 86–93.

Herbst, William and George E. Assousa. "Supernovas and Star Formation." *Scientific American* Vol. 241 (August 1979): 138–145.

Holland, H. D., B. Lazar, and M. McCaffrey. "Evolution of the atmosphere and oceans." *Nature* Vol. 320 (March 6, 1986): 27–33.

Schopf, J. William and Bonnie M. Parker. "Early Archean (3.3-Billion to 3.5-Billion-Year-Old) Microfossils from Warrawoona Group, Australia." *Science* Vol. 237 (July 3, 1987): 70–72.

Toon, Owen B. and Steve Olson. "The Warm Earth." *Science 85* (October 1985): 50–57.

Wright, I. P. "Ices in the Solar System." *Nature* Vol. 308 (April 19, 1984): 692.

DISCOVERY OF ICE

Bailey, Richard H. *Glacier*. Time-Life Books. 1982.

Chorlton, Windsor. *Ice Ages*. Time-Life Books, 1983.

Matthews, Samuel W. "Ice On The World." *National Geographic* (January 1987): 79–103.

Mintz, Leigh W. *Historical Geology: The Science of a Dynamic Earth.* Charles Merrill, 1972.

Stokes, W. Lee. *Essentials of Earth History.* Prentice-Hall, 1982.

Weiner, Jonathan. *Planet Earth.* Bantam Books, 1986.

Wilcoxson, Kent H. *Chains of Fire: The Story of Volcanoes.* Chilton, 1966.

Herbet, Wally. "Did Peary Reach the Pole." *National Geographic* Vol. 174 (September, 1988): pp. 387–413.

ANCIENT ICE AGES

Brock, Thomas D. "Precambrian evolution." *Nature* Vol. 288 (November 20, 1980): 214–215.

Kerr, Richard A. "How to Make a Warm Cretaceous Climate." *Science* Vol. 223 (February 17, 1984): 677–678.

Monastersky, Richard. "Warm Cretaceous Earth: Don't hold the ice." *Science News* Vol. 133 (June 18, 1988): 391.

——————. "Amber yields samples of ancient air." *Science News* Vol. 132 (November 7, 1987): 293.

Officer, Charles B. and Charles L. Drake. "The Cretaceous-Tertiary Transition." *Science* Vol. 219 (March 25, 1983): 1383–1390.

Rossow, William B., Ann Henderson-Sellers, and Stephen K. Weinreich. "Cloud Feedback: A Stabilizing Effect for the Early Earth?" *Science* Vol. 217 (September 24, 1982): 1245–1247.

Savin, Samuel M. "Pre-Pleistocene Climates." *Nature* Vol. 286 (August 7, 1980): 553–554.

Thomsen, D. E. "Weak Sun Blamed on WIMPS." *Science News* Vol. 128 (July 13, 1985): 23.

PLEISTOCENE GLACIATION

Aigner, Jean S. "Early Arctic Settlements in North America." *Scientific American* Vol. 253 (November 1985): 160–169.

Andrews, J. T. "Short ice age 230,000 years ago?" *Nature* Vol. 303 (May 5, 1983): 21–22.

Bahn, Paul G. "Ice Age drawings on open rock faces in the Pyrenees." *Nature* Vol. 313 (February 14, 1985): 530–531.

Bower, Bruce. "Extinctions on Ice." *Science News* Vol. 132 (October 31, 1987): 284–285.

Funnell, Brian. "Onset of the ice age in the North Atlantic." *Nature* Vol. 307 (February 16, 1984): 597.

Lewin, Roger. "Domino Effect Invoked in Ice Age Extinctions." *Science* Vol. 238 (December 11, 1987): 1509–1510.

——————. "What Killed the Giant Mammals." *Science* Vol. 221 (September 9, 1983): 1036–1037.

Lorius, C., et al. "A 150,000-year climatic record from Antarctic ice." *Nature* Vol. 316 (August 15, 1985): 591–595.

Moore, Peter D. "Clues to past climate in river sediment." *Nature* Vol. 308 (March 22, 1984): 316.

THE HOLOCENE INTERGLACIAL

Barnes-Svarney, Patricia. "Beyond the ice sheet." *Earth Science* Vol. 39 (Summer 1986): 18–19.

Bower, Bruce. "Recasting Plaster in Late Stone Age." *Science News* Vol. 134 (October 1, 1988): 213.

COHMAP Members. "Climatic Changes of the Last 18,000 Years: Observations and Model Simulations." *Science* Vol. 241 (August 26, 1988): 1043–1051.

Diamond, Jared. "The Worst Mistake in the History of the Human Race." *Discover* Vol. 8 (May 1987): 64–66.

Hansen, J., V. Gornitz, S. Lebedeff, and E. Moore. "Global Mean Sea Level: Indicator of Climatic Change?" *Science* Vol. 219 (February 25, 1983): 996–997.

Kerr, Richard A. "Climate Since the Ice Began to Melt." *Science* Vol. 226 (October 19, 1984): 326–327.

——————. "An Early Glacial Two-Step?" *Science* Vol. 221 (July 8, 1983): 143–144.

Lewin, Roger. "Modern Human Origins Under Close Scrutiny." *Science* Vol. 239 (March 11, 1988): 1240–1241.

——————. "A Revolution of Ideas in Agricultural Origins." *Science* Vol. 240 (May 20, 1988): 984–986.

Moore, Peter. "Late-glacial climatic changes." *Nature* Vol. 291 (June 4, 1981): 380.

Overpeck, Jonathan T., et al. "Climate change in circum-North Atlantic region during last deglaciation." *Nature* Vol. 338 (April 13, 1989: 553–556.

TERRESTRIAL CAUSES OF GLACIATION

Berner, Robert A. and Antonio C. Lasaga. "Modeling the Geochemical Carbon Cycle." *Scientific American* Vol. 260 (March 1989): 74–81.

Bonatti, Enrico. "The Rifting of Continents." *Scientific American* Vol. 256 (March 1987): 97–103.

Fisher, Arthur. "What flips Earth's field." *Popular Science* Vol. 232 (January 1988): 71–74.

Jacobs, J. A. "What triggers reversals of the Earth's magnetic field." *Nature* Vol. 309 (May 10, 1984): 115.

Kerr, Richard A. "New Way to Switch Earth Between Hot and Cold." *Science* Vol. 243 (January 27, 1989): 480.

LaMarche, Valmore C., Jr. and Katherine K. Hirschboeck. "Frost rings in trees as records of major volcanic eruptions." *Nature* Vol. 307 (January 12, 1984): 121–126.

Monastersky, Richard. "The Whole-Earth Syndrome." *Science News* Vol. 133 (June 11, 1988): 378–380.

Morris, Simon Conway. "Polar forests of the past." *Nature* Vol. 313 (February 28, 1985): 739.

Stommel, Henry and Elizabeth Stommel. "The Year Without a Summer." *Scientific American* Vol. 240 (June 1979): 176–186.

Stothers, Richard B. and Michael R. Rampino. "Historic Volcanism, European Dry Fogs, and Greenland Acid Precipitation, 1500 B.C. to A.D. 1500." *Science* Vol. 222 (October 28, 1983): 411–412.

Stothers, Richard B. "The Great Tambora Eruption in 1815 and its Aftermath." *Science* Vol. 224 (June 15, 1984): 1191–1197.

Valet, Jean-Pierre, Carlo Laj, and Piotr Tucholka. "Volcanic record of reversal." *Nature* Vol. 316 (July 18, 1985): 217–218.

Weisburd, Stefi. "Forests made the world frigid?" *Science News* Vol. 131 (January 3, 1987): 9.

CELESTIAL CAUSES OF GLACIATION

Campbell, Philip. "New data upset ice age theories of glaciation." *Nature* Vol. 307 (February 23, 1984): 688–689.

Cordell, Bruce M. "Mars, Earth, and Ice." *Sky & Telescope* Vol. 72 (July 1986): 17–22.

Covey, Curt. "The Earth's Orbit and the Ice Ages." *Scientific American* Vol. 250 (February 1984): 58–66.

Fodor, R. V. "Explaining the Ice Ages." *Weatherwise* Vol. 35 (June 1982): 109–114.

Kerr, Richard A. "Milankovitch Climate Cycles Through the Ages." *Science* Vol. 235 (February 27, 1987): 973–974.

Kunzig, Robert. "Ice Cycles." *Discover* Vol. 10 (May 1989): 74–79.

Leibacher, John W., et al. "Helioseismology." *Scientific American* Vol. 253 (September 1985): 48–57.

Lo Presto, James Charles. "Looking Inside the Sun." *Astronomy* Vol. 17 (March 1989): 22–30.

Monastersky, Richard. "Ice Age Insights." *Science News* Vol. 134 (September 17, 1988): 184–186.

Parecesco, Francesco, and Stuart Bowyer. "The Sun and the Interstellar Medium." *Scientific American* Vol. 255 (September 1986): 93–99.

Williams, George E. "The Solar Cycle in Precambrian Time." *Scientific American* Vol. 255 (August 1986): 88–96.

Weneser, Joseph and Gerhart Friedlander. "Solar Neutrinos: Questions and Hypotheses." *Science* Vol. 235 (February 13, 1987): 755–759.

Wolfendale, Arnold. "A supernova for a neighbor?" *Nature* Vol. 319 (January 9, 1986): 99.

MASS EXTINCTION

Benton, Michael J. "Interpretations of mass extinction." *Nature* Vol. 314 (April 11, 1986): 496–497.

Corliss, Bruce H., et al. "The Eocene/Oligocene Boundary Event in the Deep Sea." *Science* Vol. 226 (November 16, 1984): 806–810.

Crowley, Thomas J. and Gerald R. North. "Abrupt Climate Change and Extinction Events in Earth History." *Science* Vol. 240 (May 20, 1988): 996–1001.

Jablonski, David. "Background and Mass Extinction: The Alternation of Macro-evolutionary Regimes." *Science* Vol. 231 (January 10, 1986): 129–132.

Hallam, Anthony. "End-Cretaceous Mass Extinction Event: Argument for Terrestrial Causation." *Science* Vol. 238 (November 27, 1987): 1237–1241.

Lewin, Roger. "Extinctions and the History of Life." *Science* Vol. 221 (September 2, 1983): 935–937.

_____. "Mass Extinctions Select Different Victims." *Science* Vol. 231 (January 17, 1986): 219–220.

Maddox, John. "Extinction by catastrophe?" *Nature* Vol. 308 (April 19, 1984): 685.

Raup, David M. and J. John Sepkoski, Jr. "Periodic Extinction of Families and Genera." *Science* Vol. 231 (February 21, 1986): 833–836.

Raup, David M. "Biological Extinction in Earth History." *Science* Vol. 231 (March 28, 1986): 1528–1533.

Russell, Dale A. "The Mass Extinctions of the Late Mesozoic." *Scientific American* Vol. 256 (January 1982): 58–65.

Stanley, Steven M. "Mass Extinctions in the Ocean." *Scientific American* Vol. 250 (June 1984): 64–72.

Weisburd, Stefi. "Extinction Wars." *Science News* Vol. 129 (February 1, 1986): 75–77.

Wolbach, Wendy S., Roy S. Lewis, and Edward Anders. "Cretaceous Extinctions: Evidence for Wildfires and Search for Meteorite Material." *Science* Vol. 230 (October 11, 1985): 167–169.

THE ICE MAKER

Beard, Jonathan. "Glaciers on the run." *Science 85* Vol. 6 (February 1985): 84.

Bowen, D. Q. "Antarctic ice surges and theories of glaciation." *Nature* Vol. 283 (February 14, 1980): 619–620.

Crane, Robert G. "Remote Sensing and Polar Climate." *Earth and Mineral Sciences* Vol. 55 (Spring 1986): 38–41.

Kerr, Richard A. "Ice Cap of 30 Million Years Ago Detected." *Science* Vol. 224 (April 13, 1984): 141–142.

_____. "New Evidence Fuels Antarctic Ice Debate." *Science* Vol. 216 (May 28, 1982): 973–974.

_____. "Ocean Drilling Details Steps to an Icy World." *Science* Vol. 236 (May 22, 1987): 912–913.

Mollenhauer, Erik and George Bartunek. "Glacier on the move." *Earth Science* Vol. 41 (Spring 1988): 21–23.

Peltier, W. R. "Global Sea Level and Earth Rotation." *Science* Vol. 240 (May 13, 1988): 895–900.

Penkett, S. A. "Implications of Arctic air pollution." *Nature* Vol. 311 (September 27, 1984): 299.

Radok, Uwe. "The Antarctic Ice." *Scientific American* Vol. 253 (August 1985): 98–106.

Weisburd, Stefi. "Halos of Stone." *Science News* Vol. 127 (January 19, 1985): 42–44.

Zwally, H. Jay, C. L. Parkinson, and J. C. Comiso. "Variability of Antarctic Sea Ice and Changes in Carbon Dioxide." *Science* Vol. 220. (June 3, 1983): 1005–1012.

THE GREENHOUSE EFFECT

Booth, William. "Johnny Appleseed and the Greenhouse." *Science* Vol. 242 (October 7, 1988): 19–20.

Broecker, Wallace S. "Carbon dioxide circulation through ocean and atmosphere." *Nature* Vol. 308 (April 12, 1984): 602.

Colinvaux, Paul A. "The Past and Future Amazon." *Scientific American* Vol. 260 (May 1989): 102–108.

Idso, S. B., Industrial age leading to the greening of the earth." *Nature* Vol. 320 (March 6, 1986): 22.

Kerr, Richard A. "Carbon Dioxide and the Control of the Ice Ages." *Science* Vol. 223 (March 9, 1984): 1053–1054.

_____. "Linking Earth, Ocean, and Air at the AGU." *Science* Vol. 239 (January 15, 1988: 259–260.

Monastersky, Richard. "Looking for Mr. Greenhouse." *Science News* Vol. 135 (April 8, 1989): 216–221.

_____. "Global Change: The Scientific Challenge." *Science News* Vol. 135 (April 15, 1989): 232–235.

Revelle, Roger. "Carbon Dioxide and World Climate." *Scientific American* Vol. 247 (August 1982): 35–43.

Revkin, Andrew C. "Endless Summer: Living With the Greenhouse Effect." *Discover* Vol. 9 (October 1988): 50–61.

Roberts, Leslie. "Is There Life After Climate Change?" *Science* Vol. 242 (November 18, 1988): 1010–1013.

Schneider, Stephen H. "Climate Modeling." *Scientific American* Vol. 256 (May 1987): 72–80.

Wigley, T. M. L., P. D. Jones, and P. M. Kelly. "Scenario for a warm, high CO_2 world." *Nature* Vol. 283 (January 3, 1980): 17–21.

Woodwell, George M. "Contribution to Atmospheric Carbon Dioxide." *Science* Vol. 222 (December 9, 1983): 1081–1085.

_____. "The Carbon Dioxide Question." *Scientific American* Vol. 238 (January 1978): 34–43.

Zwally, Jay H., C. L. Parkinson, and J. C. Comiso. "Variability of Arctic Sea Ice and Changes in Carbon Dioxide." *Science* Vol. 220 (June 3, 1983): 1005–1012.

Index

About the Author

Jon Erickson has written several books on earth science for TAB. He holds an advanced degree in natural science and has worked as a geologist for major oil and mining companies and as an engineer for an aerospace company. He presently writes full-time as an independent geologist.

TAB Books by the Author:
- *Volcanoes and Earthquakes*   (No. 2842)
- *Violent Storms*   (No. 2942)
- *The Mysterious Oceans*   (No. 3042)
- *The Living Earth*   (No. 3142)
- *Exploring Earth from Space*   (No. 3242)
- *Ice Ages: Past and Future*   (No. 3463)
- *Greenhouse Earth: Tomorrow's Disaster Today*   (No. 3471)

Other Bestsellers of Related Interest

SUPERCONDUCTIVITY—The Threshold of a New Technology—Jonathan L. Mayo

Superconductivity is generating an excitement not seen in the scientific world for decades! Experts are predicting advances in state-of-the-art technology that will make most exisiting electrical and electronic technologies obsolete! This book is one of the most complete and thorough introductions to a multifaceted phenomenon that covers the full spectrum of superconductivity and superconductive technology. 160 pages, 58 illustrations. Book No. 3022, $12.95 paperback, $18.95 hardcover

HOW TO MAKE A MINIATURE ZOO —3rd Edition—Vinson Brown

"A strong respect for nature is echoed throughout the text, making this title an indispensable guide for the serious student of nature."   **—ALA booklist**

With this book, children of all ages will learn how to collect and care for insects, fish, amphibians, small mammals, and birds. The miniature zoos Brown describes range from a table top in the corner of a room to a back yard filled with tall wire aviaries and freshwater pools. By following the author's tips for scientifically observing animals and their ways—in zoos or in the wild—children will gain a deeper appreciation for animal life. 256 pages, 61 illustrations. Book No. 3206, $7.95 paperback, $16.95 hardcover

THE LIVING EARTH: The Coevolution of the Planet and Life—Jon Erickson

The latest in TAB's Discovering Earth Science Series, this book explores the origin and evolution of life, including the geological and biological areas. The author covers the ice ages, the geological eras, the beginnings of life, dinosaurs, plants, and more. Discussions of extinction and the destruction of live through erosion, polution, and other hazards are also presented. This is your opportunity to discover the variety and complexity of the Earth's biosphere. 208 pages, Fully illustrated. Book No. 3142, $14.95 paperback, $22.95 hardcover

UNDERSTANDING MAGNETISM: Magnets, Electromagnets and Superconducting Magnets —Robert Wood

Explore the mysteries of magnetic and electromagnetic phenomena. This book bridges the information gap between children's books on magnets and the physicist's technical manuals. Written in an easy-to-follow manner, *Understanding Magnetism*, examines the world of magnetic phenomena and its relationship to electricity. Thirteen illustrated experiments are provided to give you hands-on understanding of magnetic fields. 176 pages, 138 illustrations. Book No. 2772, $10.95 paperback, $17.95 hardcover

101 OPTOELECTRONIC PROJECTS —Delton T. Horn

Discover the broad range of practical applications for optoelectronic devices! Here's a storehouse of practical optoelectronic projects just waiting to be put to use! Horn features 101 new projects including: power circuits, control circuits, sound circuits, flasher circuits, display circuits, game circuits, and many other fascinating projects! This book offers you an opportunity to make a hands-on investigation of the practical potential of optoelectronic devices. 240 pages, 273 illustrations. Book No. 3205, $16.95 paperback, $24.95 hardcover

BOTANY: 49 Science Fair Projects —Robert L. Bonnet and G. Daniel Keen

A rich source of project ideas for teachers, parents, and youth leaders, *Botany* introduces children ages 8 through 13 to the wonder and complexity of the natural world through worthwhile, and often environmentally timely, experimentation. Projects are grouped categorically under plant germination, photosynthesis, hydroponics, plant tropism, plant cells, seedless plants, and plant dispersal. Each experiment contains a subject overview, materials list, problem identification, hypothesis, procedures and further research suggestions. Numerous illustrations and tables are included. 176 pages, 149 illustrations. Book No. 3277, $9.95 paperback, $16.95 hardcover

Other Bestsellers of Related Interest

THROUGH THE TELESCOPE: A Guide for the Amateur Astronomer—Michael R. Porcellino

Through the Telescope is an open invitation to explore our universe. This book and an amateur astronomical telescope are all you need to meet the multitude of stars, nebulae, and deep-sky objects that can be seen on a dark, clear night. Porcellino guides you on a tour of the Moon, where you'll visit craters, mountains and rilles, and learn to identify them by their unique features. Next, you'll move out to the satellites of Jupiter, the rings of Saturn, and even the Sun. 352 pages, 217 illustrations. Book No. 3159, $18.95 paperback, $26.95 hardcover

STUDIES IN STARLIGHT: Understanding Our Universe—Charles J. Caes

Even those with only limited exposure to electromagnetic concepts will find this book engrossing and understandable. Pictures and prose relate the histories of efforts made to understand the mysteries of our universe by ancient, medieval, and modern civilizations. Caes tells a tale that inspires wonder, validates theories, and dispels myths. 240 pages, 77 illustrations. Book No. 2946, $12.95 paperback, $18.95 hardcover

THE MYSTERIOUS OCEANS—Jon Erickson

Explores far below the foamy crest and delve into the wonders of the sea—its forces, its predators, its role in the food chain, its mountains, and much more. The author covers topics of oceanography, geology, meteorology, and marine biology. 208 pages, 169 illustrations. Book No. 3042, $14.95 paperback, $22.95 hardcover

VIOLENT STORMS—Jon Erickson

This book provides up-to-date information on recurring atmospheric disturbances. The internal and external mechanisms that cause weather on the Earth and the way these forces come together to produce our climate are examined. Many photographs, line drawings, and tables, as well as a complete glossary make this engrossing book informative, entertaining and easy to read. 240 pages, 190 illustrations. Book No. 2942, $16.95 paperback, $24.95 hardcover

VOLCANOES & EARTHQUAKES—Jon Erickson

The theory of Earth's creation through dynamic and destructive compulsion—and how those same energies continue to affect our planet—is examined in this compelling book. Discover how scientists predict catastrophes and the ways in which these violent events actually benefit the Earth. 304 pages, 71 illustrations. Book No. 2842, $15.95 paperback, $22.95 hardcover

LIGHT, LASERS AND OPTICS—John H. Mauldin

A fascinating introduction to the science and technology of modern optics. Broad enough to appeal to the general science enthusiast, yet technically specific enough for the experienced electronics hobbyist, this book fully explains the science of optics. You'll explore: everyday observations on light, the theory and physics of light and atoms, computing with light, optical information storage, and many other related subjects! *Light, Lasers and Optics* is extremely well illustrated with over 200 line drawings. 240 pages, 205 illustrations. Book No. 3038, $14.95 paperback, $22.95 hardcover

Other Bestsellers of Related Interest

EXPLORING EARTH FROM SPACE—Jon Erickson

Learn how orbiting satellites are used to explore our planet. Geophysicist Jon Erickson covers the technology—how satellite images are collected and processed—and describes how this technology is applied in weather forecasting, land-use planning, geologic mapping, mineral exploration, agriculture, disaster control and more. 207 pages, 157 illustrations. Book No. 3242, $15.95 paperback, $23.95 hardcover

Prices Subject to Change Without Notice.

Look for These and Other TAB Books at Your Local Bookstore

To Order Call Toll Free 1-800-822-8158
(in PA and AK call 717-794-2191)

or write to TAB BOOKS , Blue Ridge Summit, PA 17294-0840.

Title	Product No.	Quantity	Price

☐ Check or money order made payable to TAB BOOKS

Charge my ☐ VISA ☐ MasterCard ☐ American Express

Acct. No. _____ Exp. _____

Signature: _____

Name: _____

City: _____

State: _____ Zip: _____

Subtotal $ _____

Postage and Handling
($3.00 in U.S., $5.00 outside U.S.) $ _____

Please add appropriate local and state sales tax $ _____

TOTAL $ _____

TAB BOOKS catalog free with purchase; otherwise send $1.00 in check or money order and receive $1.00 credit on your next purchase.

Orders outside U.S. must pay with international money order in U.S. dollars.

TAB Guarantee: If for any reason you are not satisfied with the book(s) you order, simply return it (them) within 15 days and receive a full refund.
BC